Basic Statistics

Basic Statistics

A Primer for the Biomedical Sciences

THIRD EDITION

OLIVE JEAN DUNN

VIRGINIA A. CLARK

A Wiley–Interscience Publication
JOHN WILEY & SONS, INC.
New York • Chichester • Weinheim • Brisbane • Singapore • Toronto

This book is printed on acid-free paper. ♾

Published simultaneously in Canada.

For ordering and customer service, call 1-800-CALL-WILEY.

Library of Congress Cataloging-in-Publication Data:
Dunn, Olive Jean.
Basic statistics: a primer for the biomedical sciences / Olive Jean Dunn and Virginia A. Clark. – 3rd ed.
p.cm. – (Wiley series in probability and statistics. Texts and references section)
"Published simultaneously in Canada."
Includes bibliographical references and index.
ISBN 0-471-35422-8 (cloth: alk. paper)
1. Medical statistics. 2. Biometry. I. Clark, Virginia, 1928– . II. Title. III. Series.
RA409.D87 2000
519.5′02461–dc21 00-026009

Printed in the United States of America.
10 9 8 7 6 5 4 3 2 1

Contents

Preface to the Third Edition

This book was designed to serve as a textbook for a one-semester course in biostatistics for students in the biomedical fields. We hope it will be useful to physicians, nurses and public health workers who become involved with research projects and wish to understand the basic concepts of statistics.

The mathematics level has been deliberately kept low; high school algebra should be sufficient for understanding the text.

The scope of statistical applications and the range of professionals in the biomedical fields have broadened since the earlier editions of this book. Therefore, material has been added to relate the discussion of biostatistics to the types of studies commonly performed by health professionals. More emphasis has been placed on data screening and graphical interpretation of the information. An introductory chapter on survival analysis has been included at the end of the text. Some computational methods have been shortened or deleted because of the widespread availability of computer-based statistical programs.

A short list of references is included at the end of each chapter. The use of the list does not necessitate knowledge of higher mathematics to obtain information on the specific topic referenced. Some excellent texts are undoubtedly missing; the ones included should provide a starting point for a reader who wishes additional information.

For the instructor who is concerned about covering the text in a single semester or quarter, some parts of the book could be omitted without resulting in difficulties understanding later chapters. Chapters 10 and 12 could be omitted in their entirety. The explanation of transformations in Sections 5.5 and 11.5 could be omitted. Sections 7.5, 8.7, 8.8, and 9.2 could also be omitted.

OLIVE JEAN DUNN
VIRGINIA A. CLARK

Professors Emerita
Biostatistics and Biomathematics
University of California, Los Angeles

Basic Statistics

CHAPTER ONE

Introduction

Statistical methods have been developed to help in understanding data and to assist in making decisions when uncertainty exists. Biostatistics is the use of statistics applied to biological problems and to medicine. In this book, the examples will be given using biomedical and public health data. Bio is taken from the Greek word bios meaning life so that actually biology includes numerous fields such as ecology, fisheries, and agriculture, but examples will not be included from those fields in this book.

In Section 1.1, we discuss the two major reasons for using statistics, in Section 1.2 we present the initial steps in the design of biomedical studies, and in Section 1.3 we define the common types of biomedical studies.

1.1 REASONS FOR STUDYING BIOSTATISTICS

Statistics has traditionally been used with two purposes in mind. The first is to summarize data so that it is readily comprehensible; this is called *descriptive statistics*. Both in the past and recently, considerable effort has gone into devising methods of describing data that are easy to interpret. The use of computers and their accompanying graphic programs have made it possible to obtain attractive and meaningful displays of data without having to employ skilled graphic artists.

The second purpose is to draw conclusions that can be applied to other cases; this is called *statistical inference*. For example, in studying the effects of a certain drug on patients with asthma, one may want to do more than describe what happened to the particular patients under study. One usually wishes to decide how to treat patients with asthma in the future.

Biostatistical techniques are now widely used both in scientific articles and in articles appearing in newspapers and on television. Learning to interpret statistical summaries of data and statistical tests enables a person to evaluate what they are reading or hearing and to decide for themselves whether or not it is sensible.

Biostatistical concepts and techniques are useful for researchers who will be doing research in medicine or public health. Here, the researchers need to know how to decide what type of study to use for their research project, how to execute the study on patients or well people, and how to evaluate the results. In small studies, they may be totally responsible for the biostatistical analysis and in large studies, they may work with a professional biostatistician. In either case, knowing the basic concepts and vocabulary of biostatistics will improve the research.

1.2 INITIAL STEPS IN DESIGNING A BIOMEDICAL STUDY

In this section, we discuss setting study objectives and making a conceptual model of the disease process. We also give two measures of evaluating how common a disease condition is.

1.2.1 Setting Objectives

"If you do not know where you are going, any road will get you there" is a disparaging remark that applies to planning a research project as well as to traveling. The *first step* in evaluating a study or in planning for one is to determine the major objective of the study. Time spent in writing a clear statement of purpose of the study will save time and energy later, especially in group projects.

In biomedical and public health studies there are two general types of objectives that underlie many of the studies that are performed. Broadly stated these objectives are

1. Determine the desirability of different treatments or preventive measures to reduce disease.
2. Assess the effects of causal or risk factors on the occurrence or progression of disease.

Here, the word disease has been use to stand for any unwanted medical or psychological condition. A risk factor may often be a genetic predisposition, an environmental exposure, or a patient's life style over which the researcher has little or no control.

The first of these objectives typically leads to what are called *experimental* studies and the second to *observational* studies. An experimental study is one where the researcher can decide on the treatment for each person or animal, and then investigate the effects of the assigned treatment. For example, in a clinical trial where one-half of the patients are assigned a new treatment for high blood pressure, their outcome could be compared with those patients assigned the current best treatment. Here, the objective is to compare the new treatment for lowering high blood pressure to the current treatment to see which treatment is more successful.

An observational study is an investigation in which the researcher has little or no control over the treatment or events. The relationship between the causal or risk factors and the outcome is studied without intervention by the researcher. For example, the proportion of smokers who develop complications after reconstructive breast surgery following mastectomy can be compared with the proportion of non-smokers who develop complications. The past history of smoking cannot be controlled by the surgeon; thus the objective is to assess the effects of this potential risk factor in an observational study.

1.2.2 Making a Conceptual Model of the Disease Process

The *second step* is to write or draw a conceptual model of the disease process under study.

If an *experimental* trial is being planned, first, the beneficial effects of the proposed treatments should be postulated; second, the time until the effects become noticeable should be estimated; and third, the duration of the effects should be estimated. Possible side effects (undesirable effects) need to be anticipated as well as when they may occur. For a more complete discussion of both the critical aspects of modeling the disease and the effects of the treatments that should be considered, see Pocock [1983] or Wooding [1994].

If an *observational* study is undertaken to determine the effects of possible risk factors on the occurrence of disease, then a conceptual model of the disease process should be made. A possible framework is provided in Figure 1.1. The purpose of this diagram is to remind the investigator of the possible causal factors (often called risk factors) that could affect the disease and when they are likely to occur. For example, some *causal events* may be genetically occurring before birth, some at birth, or later in life. There may be multiple events. Some of these events may have one time occurrences (accident, major life event, exposure to a virus, etc.) and some may be chronic events (smoking, diet, continual stress, exposure to a low-level toxic chemical, etc.).

The period between the occurrence of the event and the occurrence of the disease is called the *induction* or incubation period. This induction period could be quite short or very long. In general, the longer the induction period the more difficult it is to determine the effect of a risk factor on the occurrence of the disease. The period between the occurrence of the disease and its detection is called the *latent* period. For a broken leg, this period could be very short, but for some diseases, such as multiple sclerosis, it could continue for years.

Note that to be a risk factor the exposure or causal event has to occur in the proper time period. For example, if lung cancer was detected in a patient in October 2000

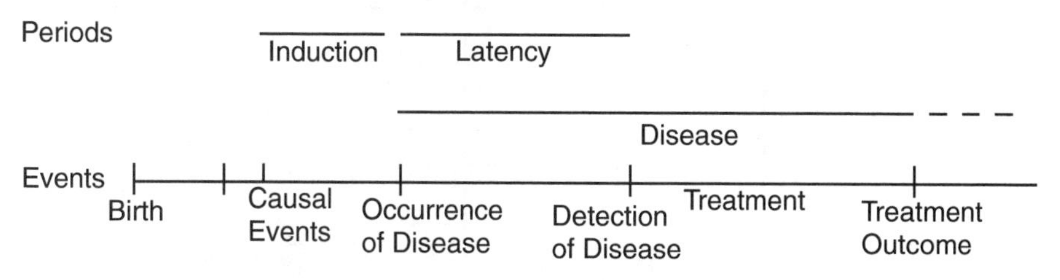

Fig. 1.1. Conceptual model of the disease process.

and the patient did not start smoking until September 2000, then smoking could not count as a risk factor for that patient since the induction period was too short.

Following detection, there is usually a *treatment* period. Following a successful treatment, the patient may become disease free (acute illness), or continue with the illness but with fewer symptoms or signs (chronic illness), or the disease could go into remission.

1.2.3 Estimating the Number of Persons with the Risk Factor or Disease

In addition to modeling the course of the disease process, it is useful in planning a study to estimate the proportion of possible subjects who have been exposed to the risk or causal factor. For example, if the risk factor is smoking, we know that a sizable proportion of the adult population currently smokes or has smoked in the past.

Finally, it helps to know the proportion of persons who already have the disease in the population under study—often called the *prevalence* of the disease. For rare diseases, even finding a sufficient number of cases to study can be difficult.

Sometimes in planning a study only newly diagnosed patients are used. In these studies, we need to estimate the number of *new* cases that will occur in the time interval allocated for patient entry into the study. This estimate is made from the incidence rate for the disease being studied. The *incidence rate* is the number of new cases occurring in a fixed time interval (usually 1 year) divided by the sum of the length of time each person in a population is at risk in the interval (see Rothman [1986] for a more complete definition). If one waits for newly diagnosed patients who can be treated at a clinic, then it is the incidence rate that is critical. For chronic diseases, it is possible to have a rather high prevalence without having a high incidence rate if the disease is not fatal.

In Figure 1.2, the results for five hypothetical adults that have been followed for 1 year are displayed. In this figure, D denotes a person being diagnosed with the disease. The vertical line indicates the time t when the prevalence study occurred. The last column gives the number of years at risk of the disease for each person. The sum of the last column is used as the denominator in computing the incidence rate.

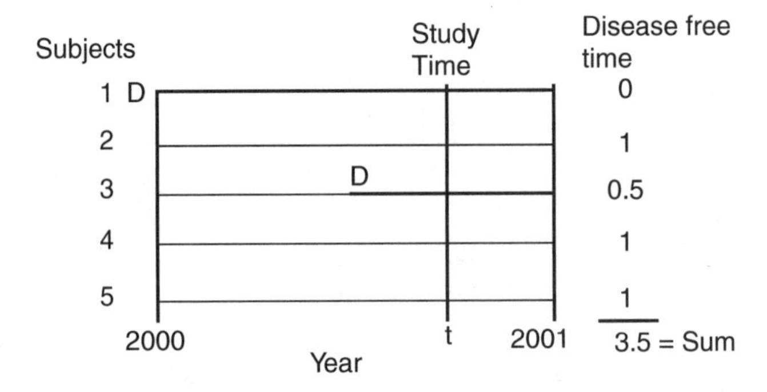

Fig. 1.2. Example of the status of five adults followed for a 1-year period.

Person 1 enters the year with the disease already known to be present. Persons 2, 4, and 5 neither have the disease before the year starts nor are diagnosed with the disease during the year. Person 3 is diagnosed with the disease one-half of the way through the year. In computing the incidence rate only one-half of the year will be counted for person 3 since they are not at risk to get the disease in the second one-half of the year. The incidence rate would be $1/3.5$. The prevalence at study time t would be two-fifths.

For rare diseases, it is common to ignore the time period that is lost due to a patient getting the disease within the study period. If this is done, then a one replaces the one-half for the disease-free time period for person 3 and the incidence rate would be one-fourth. In this case, the *incidence rate* is the number of new cases of a disease in a given time period divided by the population at risk of developing the disease.

The timing of events such as when exposure to risk factors occurs or when the disease is likely to be diagnosed, and the number of persons who have either the risk factor or the disease, limits the type of study that can be done. It is critical to have estimates of the timing and the numbers before designing a biomedical or public health study.

1.3 COMMON TYPES OF BIOMEDICAL STUDIES

In this section, a brief introduction to some of the most common types of biomedical studies is presented. Further discussion of these studies is given in subsequent chapters in this book. Here, we are concerned with when the subjects are sampled relative to the course of their disease or treatment and under what circumstances this type of study is used. The actual sampling methods are discussed in Chapter 2. (By a sample, we simply mean a subset of a selected population that the investigators wish to study.)

In terms of time, we can examine data taken at an "instant in time," we can look forward in time or we can look backward in time. The studies that theoretically take place in an instant of time are usually called *surveys* or cross-sectional studies. The types of studies that look forward in time are often called (1) *experiments*, (2) *clinical trials*, (3) *field trials*, or (4) *prospective* or panel or follow-up studies. Clinical trials and field trials are special types of experiments. Finally, biomedical studies looking backward in time are called *case/control* or case/referent studies. These studies are displayed in Figure 1.3. The arrows denote the direction of the measurement and the question marks are placed where there is uncertainty in the factor measured. The word *cause* can either be a (1) treatment or preventive measure or (2) the occurrence of a causal or risk factor. The *effect* is the outcome of the risk or causal factor or a treatment.

1.3.1 Surveys

Surveys or cross-sectional studies are used in public health and biomedical studies. In these studies, the participants are only measured once so that information on exposure

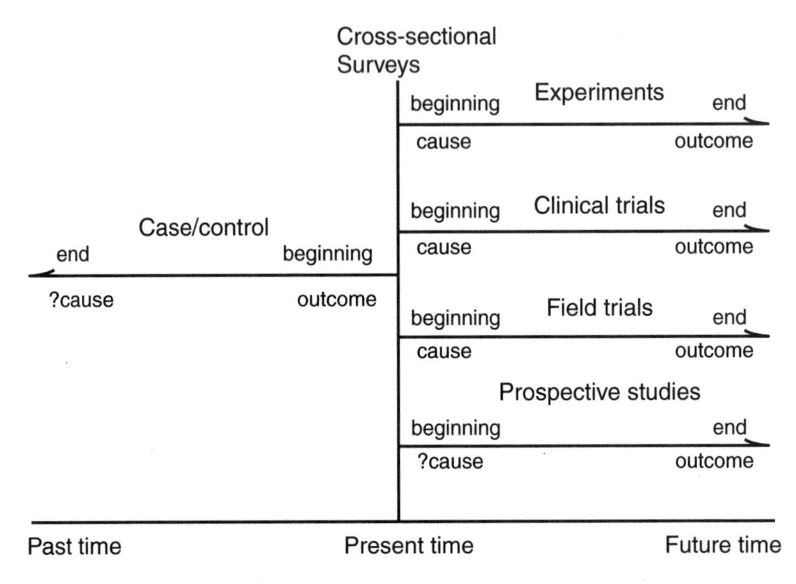

Fig. 1.3. Time relationships in biomedical studies.

or risk factors, treatments, outcomes, or other factors are all obtained at the same time. Surveys are used when both the disease (outcome factor) and the supposed risk factor (causal factor) have a high prevalence. A survey of a prevalent condition such as depression or osteoarthritis is appropriate. On the other hand, stomach cancer is a disease with a fairly low prevalence, so if a survey was taken of the general population it would have to include a huge sample just to obtain a handful of cases. Obviously, chronic conditions are easier to study with surveys than acute diseases. Sometimes surveys are done using populations that are at high risk for a given disease such as the elderly. In the public health field, surveys have been used to study the association of possible causal factors such as smoking and life style on overall health status.

One problem with surveys is that the information that can be most accurately obtained is current information on risk factors or disease conditions. It is difficult to obtain reliable and unbiased information on events that happened long ago both in terms of the event and the timing. For example, referring back to Figure 1.1, it may be difficult to determine that a causal factor occurred during the relevant time period, particularly if the causal factor occurred in a short time period.

1.3.2 Experiments

Experiments are usually simpler to interpret than surveys. Typically, a single sample is obtained and then the sample is split into different groups, often called treatment and control groups. Treatment denotes the new treatment being tested and control is either no treatment, a placebo treatment, or a standard treatment that is useful for comparison purposes. The experimental subject or unit could be an animal, person, tissue culture, and so on. Random assignment of the subject to treatment or control is the recommended practice. Discussion of how to perform a random assignment will be given in Section 2.4.3. The aim is to have the only difference between the treatment and control group be the treatment itself, not what or who is being treated.

In an experiment, the causal factor is given first and there is no question as to what the factor is since the experimenter makes that decision. The measurement of the outcome follows the treatment in time so causality can be established if the experiment is carefully designed. This is illustrated in Figure 1.3 by the absence of a question mark.

1.3.3 Clinical Trials

In clinical trials using patients, the major objectives are to test what is called efficacy (is the treatment better than the control in treating the disease) and toxicity (does the treatment have fewer side effects or unwanted effects than the control). These are complex studies to perform, which involve obtaining human subject consent, decisions on entry criteria of the patients and under what circumstances the treatment could be modified, or the clinical trial could be terminated early. Also, because the patients normally come for treatment singly over time, it may take several years to accumulate a sufficient number of patients or patients may have to be accepted at several medical centers in order to have a sufficient sample size.

Random assignment to treatment is still needed and methods of performing the assignment have been devised to fit the needs of clinical trials. If possible, it is recommended that neither the patient nor the professionals who interact with the patient or evaluate the outcome know whether the patient is getting the new treatment or the control treatment. When this can be accomplished the clinical trial is called a double-blind trial. This ideal condition is not possible in all instances. For example, it is difficult if the treatment is surgery where what was done is obvious to the patient or surgeon or if different side effects occur in a drug trial. For further information on this topic, see Mould [1998], Pocock [1983], Wooding [1994], or Spriet et al. [1994].

1.3.4 Field Trials

Field trials are performed using subjects who are not patients, and hence often require visiting the subjects in the field, which may be a school, church, or other location convenient for the subjects. An example of a field trial would be a vaccine trial. One of the largest vaccine trials was for the Salk vaccine, which had been developed to prevent polio. A vaccine or placebo was given to over 1,000,000 school children, and then the incidence of polio in the two groups was assessed. It was necessary to make the sample size large because the incidence of polio was not high. Randomization of individuals is often difficult to do in field trials because of difficulty in obtaining acceptance in the field.

1.3.5 Prospective Studies

In a typical prospective study, no treatment is assigned. There are three general types of prospective studies.

1. In *prospective trend* studies, repeated samples of different individuals can be taken at intervals over time from a dynamic population where some of the individuals in the population may change over time. Such studies have been used to study voting intentions as an election approaches.
2. In *prospective cohort* studies, repeated samples of different subjects are taken from the same cohort (group of people). Such studies have been used to determine how students change their attitudes as they progress in school.
3. In *prospective panel* studies, repeated measures are made on the same individuals over time. This is the type of prospective study most used in biomedical studies. Hereafter in this book, if the term prospective study is used it will be assumed that a prospective panel study is meant. Note that epidemiologists sometimes call panel studies cohort studies or simply prospective studies.

In the classical epidemiological prospective (panel) study, a cohort of disease-free individuals are measured for the exposure to the causal factor(s) at the beginning of the follow-up period. Then, at subsequent examinations, exposure can be remeasured and disease status(outcome) must be measured. In prospective panel studies, the subjects can be classified into two groups based on their exposure status. The critical point is that the measurement of exposure precedes the measurement of the outcome in time. There is no opportunity for the disease to have an effect on the exposure factor.

In order to use this design, it is necessary that the disease being studied have a high incidence rate in order to accumulate enough cases (persons with the disease) in a reasonable follow-up period. The disease studied must be severe enough so that its occurrence can be detected. A sufficient number of persons with exposure must be available at the start of the study. In order to obtain a sufficiently high incidence rate of the disease in the study group, high risk groups are often followed.

The follow-up period should be long enough so that the exposure could possibly cause the disease. In other words, the length of the induction and latency period must be considered. It is sometimes easier to make inferences from prospective studies if the exposures are severe and short in time. For example, a high-level exposure to radiation or toxic waste that took place in a known short-time interval would be easier to study than a low-level exposure over a considerable time period. Examples of this short-time interval would be the Chernobyl accident in Russia or the 1999 nuclear accident in Japan. However, if the exposure is chronic, then one can investigate cumulative exposure. For example, when the exposure is smoking, then often it is measured in pack years (number of years smoked times the average number of packs per day).

1.3.6 Case/Control Studies

In case/control or case/referent studies, the investigator begins with cases who already have the disease diagnosed (outcome) and looks back earlier in time for possible causes (see Schlesselman [1982]). Before discussing this type of study some mention should be made of single sample case studies (no controls without the disease used) . Here the investigator typically searches a medical record system for all the cases or patients who have a particular disease outcome in a fixed time period, say the last 2 years. Then, a search is made through the records to see if some prior exposure occurred more than would be expected considering the group of patients involved. One difficulty with this type of study is that it is difficult to evaluate the levels of the exposure factor and decide what is high or low since only cases are studied.

If both cases and controls can be studied simultaneously, then an opportunity exists to utilize a very efficient study design. A group of cases is taken and simultaneously a group of controls should be taken from the same population that the cases developed from (see Section 2.4.5 for sampling considerations). Case/control studies have the advantage of no risk to the patients, low cost, and feasibility even for rare diseases. Since most diseases are uncommon in the general population, case/control studies are often the only cost effective approach. They are not necessarily a good choice if the prevalence of the exposure factor is very rare or if there is a possibility of major differences in ascertaining the exposure factor between cases and controls.

For example, in Los Angeles County, the University of Southern California Cancer Center obtains information from hospital pathologists on essentially all newly diagnosed cases of cancer in the County. Case/control studies using these newly diagnosed cases and neighborhood controls have been successful in extending the knowledge of risk factors for several relatively rare types of cancer.

1.3.7 Other Types of Studies

There are other types of studies that are used in biomedical research such as the use of *historical controls*. For example, in surgical practice, if a surgeon decides that a new surgical technique should be used, often all subsequent patients are given the new treatment. After using the new treatment for a year or two, they may decide to compare the cases treated with the new treatment to controls treated with the previous surgical treatment in the same hospital in a time period just before the switch to the new technique. The term historical controls is used because the control patients are treated earlier in time. Inferences are more difficult to make in historical studies because other changes in the treatment or the patients may occur over time.

The topics discussed in this chapter are conceptual in nature but they can have a profound effect on the final statistical analysis. The purpose of this book is to provide information on how to analyze data obtained from these various study designs, so that after the analysis is completed, the data can be understood and inferences made from it.

PROBLEMS

1.1 List two diseases that would be possible to study with (a) a cross-sectional survey, (b) a case/control study, (c) a prospective panel study, and (d) a clinical trial.

1.2 For one of the diseases chosen in Problem 1.1 for a prospective panel study, try to make a figure comparable to Figure 1.1 for a typical patient estimating actual years between the events.

1.3 Which type of study requires (a) High incidence rate? (b) High prevalence proportion? (c) High funding? (d) Treatment under control of the investigator?

1.4 What type of diseases have high prevalence proportions relative to their incidence rate?

1.5 If you wished to study the effects of estrogen on bone density in post-menopausal women, state one precise study objective of this study. Given your precise study objective, what type of study would you recommend?

1.6 What type of study is often done when there is a low incidence rate and little funding?

REFERENCES

Mould, R. F. [1998]. *Introductory Medical Statistics*, 3rd ed., Bristol: Institute of Physics Publishing, 242–267.

Pocock, S. J. [1983]. *Clinical Trials: A Practical Approach*, New York: Wiley.

Rothman, K. J. [1986]. *Modern Epidemiology*, Boston: Little, Brown and Company, 23–34.

Schlesselman, J. J. [1982]. *Case-Control Studies*, New York: Oxford University Press, 14–38.

Spriet, A., Dupin-Spriet, T., and Simon, P. [1994]. *Methodology of Clinical Drug Trials*, Basel: Karger.

Wooding, W. M. [1994]. *Planning Pharmaceutical Clinical Trials*, New York: Wiley.

CHAPTER TWO

Populations and Samples

This chapter is an introductory chapter on sampling. In Section 2.1, we present the basic concepts of samples and populations. Section 2.2 discusses the most commonly used methods of taking samples and the reasons for using a random sample. Methods of selecting a random sample are given in Section 2.3. The characteristics of a good sampling plan and the methods commonly used in the various types of biomedical studies are described in Section 2.4.

2.1 BASIC CONCEPTS

Before presenting methods of taking samples, we define three basic concepts: samples, population, and target population. A brief discussion of the reasons for sampling is also given.

In most research work, a set of data is not primarily interesting for its own sake, but rather for the way in which it may represent some other, larger set of data. For example, a poll is taken in which 100 adults are asked whether they favor a certain measure. These 100 adults live in a small communitity that has 2500 adults living in it. The opinions of the 100 people may be of very minor importance except as an indication of the opinions prevailing among the 2500 adults. In statistical terminology, the group of 100 people is called a *sample*; the group from which the sample is taken is called the *population* (another word for the same concept is *universe*).

As an example from public health, consider a dietitian who wishes to study the diet and weight of freshmen at a certain university. Here the population is all freshmen; possibly there are 2000 of them. The dietitian can record diet histories and measure the weights of all the 2000 freshmen but may prefer to select a smaller group of freshmen (say 50 of them) and measure their diet and weights to try to learn something about the diet and weights of all 2000 freshmen. Suppose that this decision is made and measurements of the diet and weights of 50 freshmen are taken. The 50 freshman are a sample from the population of 2000 freshman.

Each freshman in the sample is sometimes called an *observational unit*. More commonly if they are people, they are referred to as individuals, cases, patients, or subjects. Populations are usually described in terms of their observational units, extent (coverage), and time. For example, for the population of freshman, the observational units are freshman, the extent is say Midwest University, and the time is Fall 2000. Careful definition of the population is essential for selection of the sample.

The information that is measured or obtained from each freshmen in the sample is called the *variables*. In this example, diet and weight are the two variables being measured.

In the two examples just given, the population from which the samples are taken is the population that the investigator wishes to learn about. But in some cases, the investigator expects that their results will apply to a larger population often called a *target population*. Consider a physician who wishes to evaluate a new treatment for a certain illness. The new treatment has been given by the physician to 100 patients who are a sample from the patients seen during the current year. The physician is not primarily interested in the effects of the treatment on these 100 patients or on the patients treated during the current year, but rather in how good the treatment might be for any patient with the same medical condition. The population of interest, then, consists of all patients who might have this particular illness and be given the new treatment. Here, the target population is a figment of the imagination; it does not exist, and indeed the sample of 100 patients who have been given this treatment may be the only ones who ever will receive it. Nevertheless, this hypothetical target population is actually the one of interest, since the physician wishes to evaluate the treatment as if it applies to patients with the same illness. We might use the following definitions: the target population is the set of patients or observational units one wishes to study; the population is what we sample from; and the sample is a subset of the population, and is the set of patients or observational units one actually does study.

Questions arise immediately. Why do we study a sample rather than the entire population? If we desire information concerning an entire population, why gather the information from just a sample? Often the population is so large that it would be virtually impossible to study the entire population; if possible, it may be too costly in time and money. If the target population consists of all possible patients suffering from a certain illness and given a certain treatment, then no matter how many patients are studied, they must still be considered to be a sample from a very much larger population.

When it is feasible to study the entire population, one may learn more about a population by making a careful study of a small sample than by taking limited measurements on the entire population. In the example of the population of 2000 freshman, the dietitian could afford the time and effort necessary to carefully collect diet histories and measure weights of 50 freshmen but not of all 2000 freshman. Accurate and detailed information on a sample may be more informative than inaccurate or limited information on the population.

2.2 DEFINITIONS OF TYPES OF SAMPLES

In this section, we define simple random samples and two other types of random samples.

2.2.1 Simple Random Samples

There are many types of samples; here we shall discuss the simplest kind of sample, which is called a *simple random sample*. A sample is called a simple random sample if it meets two criteria. First, every observational unit in the population has an equal chance of being selected. Second, the selection of one unit has no effect on the selection of another unit (all the units are selected independently). If we wish, for instance, to pick a simple random sample of 4 cards from a deck of 52 cards, one way is to shuffle the deck very thoroughly and then pick any 4 cards from the deck without looking at the faces of the cards. Here the deck of 52 is the population, and the 4 cards are the sample.

If we look at the faces of the cards and decide to pick 4 clubs, then we are not choosing a simple random sample from the population, for many possible samples of 4 cards (e.g., 2 diamonds and 2 hearts) have no chance at all of being selected. Also, if we make four separate piles of the cards; one all hearts, one all diamonds, one all spades, and one all clubs and take 1 card from each pile at random we still do not have a simple random sample from the deck of 52 cards. Each card has an equal chance of being selected, but once one heart is selected we cannot select another heart. Thus, the selection of one unit has an effect on the selection of another unit.

2.2.2 Other Types of Random Samples

It is possible to have random samples that are not simple random samples. Suppose an investigator wishes to sample senior engineering students at a particular college, and that in this college there are appreciably more male engineering students than female. If the investigator were to take a simple random sample it would be possible that very few female students would be sampled, perhaps too few to draw any conclusions concerning female students. In cases such as this, often a *stratified sampling* is used. First, the investigator would divide the list of engineering students into two subpopulations or strata, male and female. Second, separate simple random samples could be drawn from each stratum. The sample sizes of these samples could be either the same fraction from both strata (say 10%) or different fractions. If the *same* proportion of students is taken from the male and female strata, then the investigator has at least made sure that some female students will be drawn. The investigator may decide to sample a higher proportion of female students than male students in order to increase the number of females available for analysis. This latter type of sample is called a disproportionate stratified sample.

Another frequently used method of sampling is called *systematic sampling*. To obtain a systematic sample of size 50 from a population of 2000 freshmen, one begins with a list of the freshmen numbered from 1 to 2000. Next, the sampling interval $k = 2000/50$ or 40 is computed. Then, a random number between 1 and k (in this case between 1 and 40) is chosen. Note that methods for choosing random numbers are given in Section 2.3. Suppose the random number turns out to be 14. Then, the first freshman chosen is number 14. Then, every kth or 40th freshman is chosen. In this example, number 14 is the first freshman, $14 + 40 = 54$ is the second freshman chosen, $54 + 40 = 94$ is the third freshman chosen, and so on.

There are several advantages to using systematic samples. They are easy to do and readily acceptable to staff and investigators. They work well when there is a time ordering for entry of the observational units. For example, sampling every kth patient entering a clinic (after a random start) is a straightforward process. This method of sampling is also often used when sampling from production lines. Note that in these last two examples we did not have a list of all the observational units in the population. Systematic samples have the advantage of spreading the sample out evenly over the population.

Systematic sampling also has disadvantages. One of major disadvantages is a theoretical one. In Section 4.2, we will define a measure of variation called the variance. Theoretically, there is a problem in estimating the variance from systematic samples but in practice, most investigators ignore this problem and compute the variance as if a simple random sample had been taken. When a periodic trend exists in the population, it is possible to obtain a poor sample if your sampling interval corresponds to the periodic interval. For example, if you were sampling daily sales of prescription drugs at a drugstore and if sales are higher on weekends than weekdays, then if your sampling interval was 7 days you would either miss weekends or get only weekends. If it can be assumed that the population is randomly ordered, then this problem does not exist.

For a more complete discussion of these and other types of samples, see Kalton [1983], Barnett [1994], Levy and Lemeshow [1999], or for very detailed discussions Kish [1965]. Many types of samples are in common use; in this book it is always assumed that we are studying a simple random sample. That is, the formulas will only be given assuming that a simple random sample has been taken. These formulas are the ones that are in common use and are often used by investigators for a variety of types of samples.

Two other questions should be considered: Why do we want a simple random sample, and how can we obtain one?

2.2.3 Reasons for Using Simple Random Samples

The main advantage of using a simple random sample is that there are mathematical methods for these samples that enable the research worker to draw conclusions concerning the population. For random samples, we can also use similar mathematical methods to reach conclusions but the formulas are slightly more complex. For other types of samples, these mathematical tools do not apply. This means that when you want to draw conclusions from your sample to the population you sampled from, there will be no theoretical basis for your conclusions. This does not necessarily mean that the conclusions are incorrect. Many researchers try mainly to avoid allowing personal bias to affect their selection of the sample.

2.3 METHODS OF SELECTING SIMPLE RANDOM SAMPLES

Next, we will describe how to take simple random samples by simply physically drawing a small sample from a population. Random numbers will be discussed and it will be shown how they can be used to obtain a simple random sample.

2.3.1 Selection of a Small Simple Random Sample

Table 2.1 gives a set of 98 blood cholesterol measurements of untreated men aged 40–49. Each horizontal row in Table 2.1 contains 10 measurements except the last row that contains eight measurements. For illustrative purposes, these 98 blood cholesterols may be considered as a population from which we wish to select a simple random sample of size 5, say. Since the population is quite small, one way of picking a random sample from it would be to copy each measurement on a small tag. By mixing the tags well in a large container, we can draw out one tag (without looking at the writing on the tags in the container), write down the result, and then replace the tag into the container. If we follow this procedure five times, we obtain a simple random sample of five measurements.

2.3.2 Tables of Random Numbers

In larger populations, it becomes impracticable to make a tag for each member of the population, and then stir them in a container so that one must attempt in some other way to obtain a random sample. Sometimes it is possible to obtain a list of the population; then, we can easily draw a random sample by using a table of random numbers.

Table A.1 of the Appendix is a table of random digits. These single digits have been obtained by a process that is equivalent to drawing repeatedly (and immediately replacing) a tag from a box containing 10 tags, one marked with 0, one marked with 1, and so on, the last tag being marked with 9. Considerable effort goes into the production of such a table, although one's first thought might be that all that would be necessary would be to sit down and write down digits as they entered one's head. A trial of this method shows quickly that such a list cannot be called a random one: A person invariably has particular tendencies; they may write down many odd digits or repeat certain sequences too often.

To illustrate the use of such a table of random digits, a sample of size 5 will be drawn from the population of 98 blood cholesterols. First, the list of the population is made; here it is convenient to number in rows, so that number 1 is 289, number 2 is 385, number 50 is 378, and number 98 is 309. Since there are 98 measurements in

Table 2.1. Blood Cholesterol Measurements for 98 Men Aged 40–49 years (mg/100 mL)

289	385	306	278	251	287	241	224	198	287
275	301	249	288	337	263	260	228	190	282
368	291	249	300	268	283	319	284	205	294
257	256	294	253	221	241	372	339	292	294
327	195	305	253	251	229	250	348	280	378
282	311	193	242	304	270	277	312	264	262
268	251	333	300	250	234	264	291	271	284
322	381	276	205	251	270	254	299	273	252
280	411	195	256	387	241	245	325	289	306
232	293	285	250	260	316	352	309		

the population, for a sample of size 5 we need five 2 digit numbers to designate the 5 measurements being chosen. With 2 digits we can sample a population whose size is up to 99 observations. With 3 digits we could sample a population whose size is up to 999 observations. To obtain five 2-digit numbers, we first select a starting digit in Table A.1 and from the starting digit we write down the digits as they occur in normal reading order. Any procedure for selecting a starting value and proceeding is correct as long as we choose a procedure that is completely independent of the values we see in Table A.1. For an example that is easy to follow, we could select the first page from Table A.1 and select pairs of numbers starting with the 2-digit number in the upper left hand corner that has a value of 10. We proceed down the column of pairs of digits and take 37, 8, 99, and 12. But we have only 98 observations in our population, so the 99 cannot be used and the next number, or 12, should be used instead for the fourth observation. Since we need an additional number we take the next number down or 66. The final list of 5 observations to be sampled are 10, 37, 8, 12, and 66.

Referring back to Table 2.1, the cholesterol measurement corresponding to the tenth observation is 287 since we are numbering by rows. In order, the remaining 4 observations are 372 for observation 37, 224 for observation 8, 301 for observation 12, and 234 for observation 66.

Statistical programs can also be used to generate random numbers. Here the numbers will range between 0 and 1, so the decimal point should be ignored to obtain random digits.

2.3.3 Sampling with and without Replacement

The first example, in which 5 tags were drawn from a box containing 98 tags marked with blood cholesterol measurements, illustrates simple random sampling with replacement. The tags were removed from the container as they were chosen and were replaced, so that the same individual could be sampled again. This is called sampling with replacement since the tag is replaced into the container (the population). For theoretical purposes, sampling with replacement is usually assumed. In the blood cholesterol example, if we replace the tag, it is possible for the same tag to be drawn more than once. If we sample using numbers from a random number table, it is possible that the same random number will reappear so we will be sampling with replacement if we simply take all the random numbers as they appear.

If we do not replace each tag after it is drawn, then we are sampling without replacement. If we wish to sample without replacement from a random number table or from a statistical program, we will not use a random number if it is a duplicate of one already drawn.

The reason that sampling with replacement is sometimes discussed whereas sampling without replacement is actually performed is that the mathematical formulas are simpler for sampling with replacement. A moment's consideration should convince the reader that it makes little difference whether we replace or not when the sample size is small compared with the size of the population. Small is often taken to mean that the sample size is $<$ 5 or 10% of the population size. The symbol $<$ denotes less than. If the sample size is an appreciable proportion of the population, then special formulas need to be used (see Kalton [1983], Barnett [1994], Kish [1965], or other textbooks devoted to sampling) for these formulas. Sampling with replacement is not practical in most studies.

2.4 APPLICATION OF SAMPLING METHODS IN BIOMEDICAL STUDIES

In Section 2.4.1, we discuss the characteristics of a good sample. How sampling is usually done for the types of biomedical studies described in Chapter 1 is given in Section 2.4.2 to Section 2.4.5. The reasons for the use and the non-use of simple random sampling can be seen from the discussion of sampling plans used in biomedical studies.

2.4.1 Characteristics of a Good Sampling Plan

What is a good sample plan? How do we distinguish a good one from a bad one? Here, we will briefly describe the characteristics of a good sample plan.

The first criterion for a good sample is *feasibility*. The plan should be feasible in terms of what is being sampled, how it occurs, what personnel are available to take the sample, the number and location of observational units, the time schedule, and so on. Sampling schemes that are good theoretically but undesirable from a practical standpoint often are carried out so poorly that their superior theoretical advantages are outweighed by mistakes and omissions made in executing the plan.

The second criterion for a good sampling plan is one that is done in an *economical* fashion. For example, if a statistician wishes to estimate the average out of pocket expense of households for medical care, a plan should be devised that results in an accurate estimate of the average expenditure for the least cost. Cost in terms of money or personnel time is an important consideration in sampling and is often a major reason in the choice of the particular sampling plan.

A third criterion is the plan should be such that it is possible to make inferences from the sample to the population from which it was taken. For example, if we take a sample and find that 75% of the patients in the sample are taking their medication properly, can we infer that $\sim$ 75% of the patients in the population we sample from are also taking their medication properly? This has been called *measurability* (see Kish [1965]). This criterion is often in conflict with the first two. As has been discussed in this chapter, a simple random sample meets this criterion. Random samples, in general, allow us to meet this criterion but they may result in more complicated formulas. But often, as we shall see in subsequent sections, it is not feasible or economical to take simple random samples or even random samples.

2.4.2 Samples for Surveys

The size and location of the population often dictates what sampling method will be used in surveys. For example, the population could be all English or Spanish speaking noninstitutionalized adults in Los Angeles County in the Fall of 2000. Here the population is very large and spread out. In contrast, a sample could be taken of all medical school faculty at a given medical school in the Fall of 2000. This population will be much smaller in size and easily located.

In the first situation, where no listing of all adult residents in Los Angeles County exists and the adults are widely spread out, the most cost effective procedure is to subcontract to a professional survey group. Devising a good sample plan is too expensive and time consuming. Simple random samples are not used in sampling households from a large county such as Los Angeles because of both feasibility and cost considerations. Samplers employed by professional surveyers are able to use information from past surveys and are familiar with specialized sampling techniques for large populations that save considerable time and effort. The resulting sampling plan can still be random, but it is not a simple random sampling plan.

When a population is small and a list is already available, it is relatively easy to obtain a survey sample. For example, if the observational units are medical school physicians with full-time appointments, a list can undoubtedly be obtained. It is advisable to check on whether the records are up to date and accurate before beginning. A simple random sample can be taken, with random numbers obtained either from a table such as Table A.1 or from a statistical program. Systematic samples are also commonly used for surveys from small populations.

If the persons to be sampled live in a very small area, it is possible to drive or walk around the area and note on a map all the places that people live. This is more difficult than it sounds as people can live in unusual places such as garages or trailers or in locked residential areas where access to the households is virtually impossible. Sometimes a reverse directory supplemented by a tax roll or aerial photographs are helpful. Then, a sample of households can be taken. Often a systematic sample is the simplest one to use. From a sample of households, a subsample of adults can be obtained.

Another method of sampling households is to do a random digit dialing survey by telephone. The first step is to obtain the area code (3 digits) and the prefix numbers (3 digits) in the desired area. Other information on how the last 4 digits are assigned by the telephone company can be obtained from the company. The next step is to dial 4 digit random numbers with suitable prefix numbers (see Frey [1989] for a more detailed description of how to take a sample by telephone). Sometimes investigators will take random samples from telephone directories but especially in urban areas this procedure results in many household not being sampled because they have an unlisted number. An additional problem is that cell phones numbers are not listed in some telephone books and they may be operated by different companies from the regular phone system.

As the population gets smaller and the eligibility restrictions for defining who is in the population get stricter, the usual problem is obtaining a sufficient sample size. In many biomedical studies, this is a real concern. For example, if you wish to study persons who just came down with a rare disease you will tend to take anyone you can find who meets the entry criteria. This type of sample is often called a *chunk* or convenient sample. It will not make sense to a physician to go to a lot of effort to find suitable subjects, and then decide not to use some of them. Such samples do not meet the criteria of measurability discussed earlier but they may be the best available. For example, it is difficult to know if the patients with a particular disease who are treated by one physician are typical of all patients with the same disease.

In public health surveys, investigators often want to sample persons who follow procedures or life styles that are illegal or unpopular. Samples of volunteers can be recruited using suitable newspapers, television, radio, or internet web sites that are aimed at the desired population. For example, a sample of heavy marijuana smokers cannot be recruited in the same manner as the general public is recruited. The resulting samples are often chunk samples but they are the only type that is feasible and economical.

2.4.3 Samples for Experiments

In laboratory experiments, where animals, tissue cultures, or other observational units are used, the sampling method most used is chunk samples. Investigators tend to be careful where they purchase their animals and the type used for a given kind of experiment, but the focus is on how the treatment affects the outcome and how best to measure that outcome. An implicit assumption is made that the effects of the treatment are quite general and the animals or whatever is being experimented on can be thought of as replicates. If animals are used, it is recommended that they be randomly assigned to the treatment group and sometimes they are randomly assigned to cage location if there is thought to be any possibility that this location could affect the outcome. *Random assignment* can be easily accomplished using a random number table. Suppose there are two treatments called A and B. Then, the researcher can look in a random number table such as Table A.1, and assign an animal to A if the number is an even number and to B if it is an odd number. When dealing with three groups, if the random number chosen is 1, 2, or 3, the animal can be assigned to A; a random number of 4, 5, or 6 results in the animal assigned to B; and a random number of 7, 8, or 9 results in the animal assigned to C. Zeros would be ignored.

Clinical trials are experiments where the observational unit is the patient. The purpose is to test the efficacy (does it help) and the toxicity (unwanted effects) of a new treatment versus a placebo or a currently used standard treatment. Patients that are used must meet a variety of strictly drawn criteria and sign consent forms. Patients tend to come in gradually over time and so are not available in a group as are test animals. They are a sample taken at a given medical institution or institutions and within a given time period who meet the criteria for entry, and who agree to participate. They are often chunk samples.

In general, these clinical trials can be separated into two broad groups. The first group can be called independent trials that are done by individual or small groups of physicians. For example, a surgeon may decide to compare two methods of reconstructive breast surgery for women who have had a mastectomy. Patients who meet the criteria and agree to accept either operation will be randomly assigned to one of two surgical methods. Here, one recommended procedure is to make a list containing A and B in a random order where A designates one treatment and B the other. Again using odd/even numbers from the random number table to obtain a random order of the A and B treatments is advised. The A or B is put into consecutively numbered envelopes that are not opened until the time of the treatment. Thus neither the patient, receptionist, or surgeon can control which treatment the patient gets.

Assigning A and B to every other patient is not recommended. The personnel entering or evaluating them can determine the assignment procedure if they even know what one patient received.

Larger more formal studies are performed before drugs are approved for use. These clinical trials are divided into phases. Phase I trials, which are conducted on healthy volunteers, are conducted first to assess safety at various doses. If a drug is deemed safe, a Phase II trial is conducted to assess the optimum dose and efficacy in patients with the disease to be treated. If the drug passes this test it goes on to Phase III, where an experiment is done with patients randomly assigned to the new drug or the current standard treatment. Again, the treatments are given in a random order. Here, the random assignment to treatment may be concealed in an envelope that is given to the pharmacist or transmitted by computer. The pharmacist then gives the patient the prescription that matches what is stated inside the envelope. Special methods of assignment are sometimes used so that there is a equal number of patients in each treatment group (see Wooding [1994], Friedman and Furberg [1998], or Fleiss [1986]). This can either be used for the entire sample or a subset of the sample. For example, suppose it is known for the disease under study that for a small proportion of the patients neither treatment is likely to be successful due to the severity of the disease. Then, as these patients enter the study, they can be put into a special stratum and special assignment methods used to assure that each treatment group gets an equal number of them. Note that random assignment in general is not guaranteed to result in equal numbers in the two treatment groups.

Because of the scarcity of patients who meet all the entry criteria and are willing to take part in the clinical trial, chunk samples are often taken from consecutive patients that arrive for treatment.

2.4.4 Samples for Prospective Studies

Examination of prospective studies or panel studies show that several types of samples tend to occur. Many prospective studies have been done using a chunk sample of a special cohort of people who have been chosen for their high rate of cooperation both initially and at follow-up. For example, studies among employees in a stable industry such as utilities, the armed services, or religious groups such as the Seventh Day Adventists can be done with minimal loss to follow-up. The employer or group will keep track of the persons in the study and the investigator can use their records to find the study subjects.

Some studies have been done using random samples from a general population within a geographic area. One of the most noted examples is the Framingham Heart Study, where the sample was a mixture of adults who were randomly sampled and volunteers who after hearing of the study wished to participate (see Mould [1998]). In Los Angeles, a prospective study of depression was done where the sample was obtained in a random fashion by the Survey Research Center at UCLA (see Frerichs et al. [1981]). This type of study is more expensive to do and the loss to follow-up may be higher than when a cohesive and motivated chunk sample is taken. Nevertheless, this type of sample allows the researcher to make inferences to the population being sampled.

2.4.5 Samples for Case/Control Studies

In case/control studies, the investigator starts with the cases after they are diagnosed or treated. These studies are also called retrospective since the investigator is looking backward in time. This often involves taking a chunk sample at one or more institutions who have medical records that the investigator can search to find cases that meet the eligibility criteria for a particular disease. Sometimes the study is performed solely from available records. Otherwise, the investigator must contact the case, obtain their consent to enter the study, and interview or examine them further. These procedures will lessen the number of usable cases and may make the sample less representative of patients with this disease. Most investigators try to use the most recent patients as their records may be better and they are easier to find. Usually, a 100% sample of the most recent cases that are available to study are taken. Hence, the sample of cases is a chunk sample taken in a given place and time.

The best sample of controls is more difficult to define. They will be used to contrast their exposure history with that of the cases. The controls should be taken from the same population that gave rise to the cases and the sample chosen in such a way that the chance of being sampled is not dependent on the level of the major exposure variables. They should be similar to the cases in regards to past *potential exposure* during the critical risk period. Three types of controls have been commonly used and each has its advocates; hospital or clinic controls, friend controls, and neighborhood controls.

Typically, hospital controls are taken from the same hospital as the case, admitted at similar time, the same gender and close in age, but have a different disease condition. One method of doing this is to individually find one or more matches for each case. Here, you are trying to find a pseudotwin. The investigator has to be careful in what they choose to match, as it is not possible to analyze any factor that is used for matching purposes since the cases and controls have been made artificially equal in respect to that factor.

The matching controls from hospitals are sometimes found by sorting a list of potential controls on gender, age, and time of admission (or whatever matching variables are used) with the computer, and then finding matches from this sorted list. This procedure is sometimes called *caliper matching* because the investigator is trying to find a control that is close to the case on the matching variables. Alternatively, if the variables being matched on fall into two groups such as gender and age and are grouped into three age intervals (say, < 30 years old, 31–50 years old, and > 50 years old), then what is called *frequency matching* is commonly done. The symbol $>$ signifies greater than. Here, the investigator makes sure that there are an equal number of cases and controls in each gender and age category or perhaps two controls for each case.

Friend controls are chosen by asking the case (or a close relative if the case is dead) for the name and address of a friend or friends who can be contacted and used as a control. When obtaining friend controls is successful, it can be an inexpensive way of getting a matched control; it is thought to result in a control who has a life style similar to that of the case. The three problems that have been noted with this method are that (1) sometimes cases lack friends or are unwilling to name them; (2) cases tend to name friends who are richer or have a better job or are in some way more prestigious than themselves; and (3) if the critical exposure time was in the distant past, then cases may have trouble recalling friends from that time period.

Sometimes neighborhood controls are used either in and of themselves or to supplement unavailable friend controls. Neighborhood controls are thought of as suitable matches on general life style. These controls are found in the neighborhood of the case by use of reverse directories, maps, or neighborhood searches. The general rule for matched controls is to exclude the homes immediately adjacent to the case for privacy reasons. A list is made of 10–15 nearby houses following an agreed upon pattern. The interviewer visits the first household on the list and sees if an eligible control lives in the house (say same gender and within 5 years of age). If the answer is no, they go to the next house, and so on. Letters are left at the homes where no one is home in an attempt to encourage a contact. Not at home addresses should be repeatedly contacted since the investigator does not want a sample solely of stay at homes. Neighborhood controls are costly and time consuming to obtain but they are thought to be suitable controls.

It is difficult to describe the sample of controls in a straightforward way since they are chosen to match the cases. Viewed as a single sample, they certainly lack measurability. Their value lies in their relationship to the cases. For further discussion of case/control studies and how to choose controls, see Rothman [1986] or Schlesselman [1982].

As can be seen from this introduction to sampling for medical studies, the problems of feasibility and economy often lead to the use of samples that are not simple random samples. It takes careful planning to obtain samples that, if not ideal, can still be used with some level of confidence. The success of a good study rests on the base of a good sample.

PROBLEMS

2.1 Draw three simple random samples, each of size 6, from the population of blood cholesterol measurements in Table 2.1. Use the table of random digits to select the samples, and record the page, column, and row of the starting digit you use. Record the six individual measurements separately for each sample in a table. They will be used for subsequent problems in other chapters.

2.2 Draw one systematic sample of size 6 from the population of blood cholesterol measurements. Record what your starting value was.

2.3 Suppose you decide to use Table A.1 to assign patients to treatment A or B by assigning the patient to A if an even number (e) is chosen and B if odd number (o) is chosen. Starting with the numbers in the upper left hand corner of the first page of the table and reading across the first row, you would have oe eo oo eo oo—oe oe eo oo ee—oe eo oo ee oe—ee oo oe oo oo—oo eo eo eo eo. Does this appear to be a random allocation to you? How do you explain the fact that there were six odd numbers chosen in a row? If you use this allocation, will you have an equal number of patients receiving treatment A and B?

2.4 You are asked to take a sample of 25 students from a freshman medical class of 180 students in order to find out what specialty they wish to enter. Write out instructions for taking this sample that can be followed by a staffer with no sampling experience.

2.5 Use a random number generator from a statistical computer package to generate 10 random numbers.

2.6 An investigator wishes to take a sample of 160 undergraduate students from a university to assess the attitudes of undergraduates concerning the student health service. The investigator suspects that the attitudes may be different for freshmen, sophmores, juniors, and seniors since it is thought that the attitudes might change as the students get more experience with the system. The investigator would like to make estimates for each class as well as the total body of undergraduates. Choose a method of sampling and give the reasons for your choice.

2.7 You have examined last years pathology records at a hospital and have found records of 100 women who have newly diagnosed breast cancer. You are interested in studying whether or not risk factor X, which is believed to have its effect on risk in women whose cancer is diagnosed when they are < 50 years old, has had a significant effect in the population from which these women came. Describe how you would choose a suitable sample of controls.

REFERENCES

Barnett, V. [1994]. *Sample Survey: Principles and Methods*, London: Edward Arnold.

Fleiss, J. L. [1986]. *The Design and Analysis of Clinical Experiments,* New York: Wiley–Interscience, 47–51.

Frerichs, R. R., Aneshensel, C. S., and Clark, V. A. [1981]. Prevalence of depression in Los Angeles County, *Am. Epidemiol. J.,* **113**, 691–699.

Frey, J. H. [1989]. *Survey Research by Telephone,* Newbury Park, CA: Sage, 79–116.

Friedman, L. M. and Furberg, C. D. [1998]. *Fundamentals of Clinical Trials,* New York: Springer.

Kalton, G. [1983]. *Introduction to Survey Sampling,* Newbury Park, CA: Sage.

Kish, L. [1965]. *Survey Sampling,* New York: Wiley.

Levy, P. S. and Lemeshow, S. [1999]. *Sampling of Populations*, 3rd ed. New York: Wiley.

Mould, R. F. [1998]. *Introductory Medical Statistics,* 3rd ed., Bristol: Institute of Physics Publishing, 332–333.

Rothman, K. J. [1986]. *Modern Epidemiology,* Boston: Little, Brown and Company, 105–106.

Schlesselman, J. J. [1982]. *Case-Control Studies,* New York: Oxford University Press, 75–85.

Wooding, W. M. [1994]. *Planning Pharmaceutical Clinical Trials,* New York: Wiley.

CHAPTER THREE

Frequency Tables and Their Graphs

Usually, we can obtain only a rather vague impression from looking at a set of data, especially with a large data set. The complexity that the mind can grasp is limited, and some form of summarizing the data becomes necessary. In this chapter and in Chapter 4, commonly used ways of depicting data are considered. The general purpose of Chapters 3 and 4 is to show how data can be described in a form that alerts the reader to the important features of the data set. We regard such descriptions as a first step in analyzing data; frequently, it is the only step that needs to be taken. With small data sets it can be done by hand but with larger sets statistical computer programs should be used.

In this chapter, as a straightforward way of showing the meaning of tabulations and graphs, we give details of how they can be made by hand. First, methods of summarizing the data in frequency tables are illustrated in Section 3.1. Then, Section 3.2 presents ways of making histograms and frequency polygons from the results of the frequency tables. When statistical programs are used, the explanations will help in the choice of the program and also in choosing among the options. Even spreadsheet programs, now widely available for desktop and laptop computers, include capabilities adequate for much of the material discussed in this chapter.

3.1 NUMERICAL METHODS OF ORGANIZING DATA

In Table 3.1, we have an example of a moderate sized set of raw data containing hemoglobin levels for 90 high-altitude miners in grams per cubic centimeter. Without some rearrangement of the 90 observations, the values are difficult to interpret. Here in Section 3.1, we discuss making an ordered array of the data, and then show how to make a frequency table.

Table 3.1. Hemoglobin Levels of 90 High-Altitude Mine Workers (g/cc)

18.5	16.8	23.2	19.4	19.5	20.6	22.0	17.8	16.2
23.3	19.7	21.6	24.2	21.4	20.8	19.7	21.1	23.0
21.7	18.4	22.7	20.9	20.5	16.1	16.9	24.8	12.2
17.4	17.8	19.3	17.3	18.3	17.8	17.1	18.4	19.7
17.8	19.0	19.2	15.5	26.2	19.1	20.9	18.0	21.0
20.2	18.3	19.2	17.2	19.8	19.5	20.0	18.4	15.9
19.9	16.4	18.4	17.8	23.0	19.4	20.3	18.2	13.1
20.3	18.5	24.1	14.3	17.8	19.9	23.5	19.7	19.3
20.6	18.3	20.8	17.6	18.1	19.7	19.1	19.5	23.5
18.5	20.0	22.4	18.8	16.2	15.6	15.5	18.5	19.0

Table 3.2. Stem and Leaf Table of Hemoglobin Data

First Two Digits	Third Digit	Count	Cumulative
12	2	1	1
13	1	1	2
14	3	1	3
15	5659	4	7
16	842192	6	13
17	48832868818	11	24
18	554353483140425	15	39
19	970322458154977175730	21	60
20	236089568903	12	72
21	76410	5	77
22	740	3	80
23	320505	6	86
24	128	3	89
26	2	1	90

3.1.1 An Ordered Array

The simplest rearrangement of the data is an ordered array. An ordered array is an arrangement of the observations according to size from smallest to largest. It can be formed rather easily by hand for sets of data that involve no more than 3 digits by listing the first 2 digits in a column as in Table 3.2, and then quickly going through the numbers in Table 3.1 and recording the third digit in the appropriate row. Thus the first two values in Table 3.1 are 18.5 and 23.3, so 5 and 3 are entered in Table 3.2 beside 18 and 23, respectively.

Tables such as Table 3.2 are called *stem and leaf* tables; here the first 2 digits form the stem and the last digit is the leaf. By itself, the stem and leaf diagram is simple to make and provides a useful description of the data. From it, the investigator can look at the numerical values of the data and also can get an impression of how the data are distributed. From Table 3.2, one sees that the most common hemoglobin level is

Table 3.3. Ordered Array of Hemoglobin Levels of 90 High-Altitude Mine Workers (g/cc)

12.2	16.4	17.8	18.4	19.0	19.5	20.0	20.9	23.0
13.1	16.8	17.8	18.4	19.1	19.5	20.0	20.9	23.0
14.3	16.9	17.8	18.4	19.1	19.7	20.2	21.0	23.2
15.5	17.1	17.8	18.4	19.2	19.7	20.3	21.1	23.3
15.5	17.2	18.0	18.5	19.2	19.7	20.3	21.4	23.5
15.6	17.3	18.1	18.5	19.3	19.7	20.5	21.6	23.5
15.9	17.4	18.2	18.5	19.3	19.7	20.6	21.7	24.1
16.1	17.6	18.3	18.5	19.4	19.8	20.6	22.0	24.2
16.2	17.8	18.3	18.8	19.4	19.9	20.8	22.4	24.8
16.2	17.8	18.3	19.0	19.5	19.9	20.8	22.7	26.2

19 plus and that there are more values < 19.0 than > 19.9. Note that stem and leaf diagrams are a common option in statistical programs.

Stem and leaf tables are also useful in seeing whether the person who takes the data tended to round the numbers to whole numbers or halves. This would show up as an excess of 0's and 5's in the leaf column.

Table 3.3 contains the final ordered array. From the array, we can notice more about the data than we could from the data as shown in Table 3.1. For instance, we can easily see that all the observations lie between 12.2 and 26.2 g/cc, and that most of them lie between about 15 and 24 g/cc. A table such as Table 3.3 can be obtained from Table 3.2 or from Table 3.1 by using sort options from statistical or spreadsheet programs.

Table 3.3 is still cumbersome, however, and contains so much unimportant detail that we cannot easily distinguish its important properties. Alternatively, data may be presented in the form of a frequency table. In a frequency table, details are sacrificed in the hope of showing broad essentials more clearly. A frequency table can also be viewed as the first step in making a histogram (see Section 3.2.1).

3.1.2 The Frequency Table

To make a frequency table, we find the interval that includes the smallest and largest observation in the data set (here 12.2–26.2) and decide on some convenient way of dividing it into intervals called class intervals or classes. The number of observations that fall in each class interval are then counted; these numbers form a column headed *frequency*.

Table 3.4 shows a frequency table of the hemoglobin data for the 90 workers. In Table 3.4, the first class interval was chosen to be 12.0–12.9. Here, 12.0 was chosen for convenience as the starting point, and the length of each interval is 1 g/cc. The table succeeds in giving the essentials of the entire set of data in a form that is compact and can be read quickly.

The investigator who collected the data may find the frequency table as it stands quite adequate for their use, or they may for some purposes prefer using the original 90 observations. An investigator wishing to publish printed data for others to use may

Table 3.4. Frequency Table for Hemoglobin Levels of 90 High-Altitude Mine Workers (g/cc)

Class Interval	Midpoint	Frequency
12.0–12.9	12.45	1
13.0–13.9	13.45	1
14.0–14.9	14.45	1
15.0–15.9	15.45	4
16.0–16.9	16.45	6
17.0–17.9	17.45	11
18.0–18.9	18.45	15
19.0–19.9	19.45	21
20.0–20.9	20.45	12
21.0–21.9	21.45	5
22.0–22.9	22.45	3
23.0–23.9	23.45	6
24.0–24.9	24.45	3
25.0–25.9	25.45	0
26.0–26.9	26.45	1
		Sum = 90

publish a frequency table rather than the raw data unless the data set is small. It is important in either case that the table be properly labeled, with title, source, units, and so on. It is also important that the class intervals be designated in such a way that it is clear exactly which numbers are included in each class.

In Table 3.4, the designation of the class intervals is done in the most usual way. We might wonder, however, in looking at it, what happened to workers whose hemoglobin measurements were between 12.9 and 13.0 g/cc. The answer is that the measurements were made originally to the nearest 0.1 g/cc, so that there are no measurements listed between 12.9 and 13.0 g/cc. The class intervals were made to reflect the way the measurements were made; if the measurements had been made to the nearest 0.01 g/cc, the appropriate first interval would be 12.00–12.99.

When measurements are made to the nearest 0.1 g/cc, then any measurement between 11.95 and 12.05 is recorded as 12.0, and any number between 12.85 and 12.95 is written as 12.9, and so on. Thus the first class interval in Table 3.4 covers hemoglobin levels from 11.95 to 12.95. An alternate designation for the class intervals in Table 3.4 is, therefore, 11.95–12.95, 12.95–13.95, and so on, assuming that the data were recorded to the nearest 0.1. To make the table clear, the investigator should tell the reader how the data were recorded.

A third way of giving the intervals is 12.0–13.0, 13.0–14.0, and so on. This is unsatisfactory, for it is not clear whether a measurement of 13.0 goes into the first class or the second.

Table 3.4 also displays midpoints of each interval. The midpoint for the first interval is the average of 11.95 and 12.95 or $(11.95 + 12.95)/2 = 11.45$. Note that all hemoglobin levels between 11.95 and 12.95 fall in the first interval because the measurements have been made to the nearest 0.1 g/cc. Subsequent midpoints are found by adding one to the previous one since the class interval is 1 g/cc long. The midpoints are used to represent a typical value in the class interval.

A frequency table with different class intervals for the same data set is shown in Table 4.2. In Table 4.2, the first interval was chosen as 12.0–13.9, so that the class intervals are twice as long as in Table 3.4. It is apparent that the same set of data can be put into many different frequency tables by choosing different class intervals or by changing the beginning point of the first class interval.

There is no one "correct" frequency table such that all the rest are incorrect, but some are better than others in showing the important features of the set of data without keeping too much detail. Often a researcher chooses a class interval based on one used in the past; for researchers wishing to compare their data with that of others, this choice is helpful. Comparison is simpler if the two frequency distributions have the same class intervals.

Before constructing a frequency table, we examine the range of the data set in order to decide on the length and starting points of the class intervals. The *range* is defined to be the difference between the largest and the smallest observation, so that for the data in Table 3.1, the range is $26.2 - 12.2 = 14.0$. Next, an approximate number of intervals is chosen. Usually between 6 and 20 intervals are used; too much is lost in grouping the data into fewer than 6 intervals, whereas tables of > 20 intervals contain an unnecessary amount of detail. Intervals of equal length are desirable, except in special cases. To obtain equal intervals, the range is divided by the number of intervals to obtain an interval length that is then adjusted for convenience. In the example, the range of the data is $26.2 - 12.2 = 14.0$ g/cc, and if we wish about 12 equal intervals, $14.0/12 = 1.2$ g/cc. We then adjust the 1.2 to 1.0, a more convenient length, for the class interval.

The beginning point of the first class must still be decided upon. It must be at least as small as the smallest observation otherwise the choice is arbitrary, although 12.0 seems the most convenient one.

One problem with making frequency tables using computer programs is that some programs do not provide investigators with the freedom to choose the particular class intervals they desire. Many programs give only the frequencies of each distinct observation; these counts can be accumulated later or the user may be able to group their data as a separate procedure before obtaining the frequencies. Note that the words *frequency table* in many computer programs refers to a type of table discussed in Chapter 9.

3.1.3 Relative Frequency Tables

If the numbers in the frequency table are expressed as proportions of the total number in the set, the table is often somewhat easier to interpret. These proportions are computed by dividing the frequencies in each class interval by the total sample size. Often these proportions are converted to percentages by multiplying by 100. The table may then be called a table of relative frequencies, or it may still be called a frequency distribution or a frequency table. Relative frequency tables are especially helpful in

Table 3.5. Frequency Table of Hemoglobin Levels for 122 Low-Altitude Mine Workers Illustrating Relative Frequency and Cumulative Frequency

Hemoglobin Level (g/cc)	Frequency	Relative Frequency (%)	Cumulative Frequency (%)
11.0–11.9	6	4.9	4.9
12.0–12.9	21	17.2	22.1
13.0–13.9	29	23.8	45.9
14.0–14.9	43	35.2	81.1
15.0–15.9	19	15.6	96.7
16.0–16.9	3	2.5	99.2
17.0–17.9	1	0.8	100.0
Sum	122	100.0	

comparing two or more sets of data when the sample sizes in the two data sets are unequal.

Table 3.5 gives a frequency table showing the hemoglobin level for 122 low-altitude miners. Class intervals for Table 3.5 are of length 1.0 g/cc, as in Table 3.4. Relative frequencies reported in percents have been calculated for the results for low-altitude miners. For example, 4.9% has been calculated by dividing 6 by 122, the total sample size, and then multiplying by 100 to obtain percents.

The last column in Table 3.5 gives cumulative frequency in percent; this is the percentage of the low-altitude miners who had hemoglobin levels below the upper limit of each class interval. The figures are obtained by adding the percentages in the relative frequency column. For example, the first two relative frequencies in the table for low-altitude miners are 4.9 and 17.2; these add to 22.1, the percentage found opposite the second class interval under cumulative frequency. This says that 22.1% of the mine workers had hemoglobin levels measuring < 12.95 g/cc.

3.2 GRAPHS

The purpose of a graph is the same as that of a table; to show the essentials of a set of data visually so that they may be quickly and easily understood. In gathering data into any type of table, some of the details are lost in achieving the objective of showing the main features. In making a graph, we draw a picture of the situation, and we may lose even more of the fine details. A well-done graph is usually easier to read and interpret than the table.

Two ways of drawing a graph from a frequency table are given here: the histogram and the frequency polygon. In Section 4.6, some graphs will be described that can be used to display measures of the center and spread of a set of observations.

3.2.1 Histograms: Equal Class Intervals

When the class intervals in the frequency distribution are equal, a histogram can be drawn from it directly, using the frequencies, proportions, or percentages. Graph paper is helpful, though not essential. Two lines, one horizontal, the other vertical, are all that is needed; one line is called the horizontal axis and the other the vertical axis. A suitable scale is then marked on each axis. The horizontal scale must be such that all the class intervals can be marked on it; it need not begin at zero. The vertical scale should begin at the point where the axes cross. If the histogram is to be plotted with frequencies, the scale must be chosen so that all the frequencies can be plotted. With proportions or percentages, the vertical scale must be chosen so that the largest one can be plotted. There is a wide leeway in choosing the scales for the two axes; they could be chosen in such a way that the histogram looks tall and thin or so that it looks flat and wide; both extremes should be avoided.

After choosing the scale, class intervals are marked on the horizontal scale, and then a vertical bar is erected over each class interval. The height of each bar is made to equal the frequency, proportion, or percentage in that class.

After drawing in the bars, all that remains to be done is to make sure that the histogram is self-explanatory. This is important even when the graph is for one's own use rather than for publication. A title and the source should be given. On each axis, the scale should be marked and labeled. Figure 3.1 shows two histograms drawn from the data in Table 3.5 and relative frequencies computed for the high-altitude mine workers using Table 3.4. Percentages are plotted on the vertical axis. In Figure 3.1, the numerical values of the end points of the class intervals are marked on the horizontal axis. Alternatively, many programs mark the midpoints of the class intervals so as to indicate a typical value for that interval. Histograms may also be plotted using frequencies on the vertical axis, but percentages are preferable for purposes of comparison when the sample sizes are unequal.

In this example, the histograms clearly show higher hemoglobin levels for the high-altitude mine workers than for the low-altitude ones. Also, there is a greater range between the highest and lowest values for the high-altitude workers.

Almost all statistical packages and many spreadsheet programs include procedures for making histograms. Some programs allow the user to specify the length of the class interval and the minimum value, and so on, and others simply draw a histogram without allowing the user to specify where it starts or the width of the intervals. Many additional options are often available to make the histogram more attractive and to assist in assessing how the data are distributed.

For small sample sizes and for histograms not for publication, investigators sometimes draw quick histograms by placing an X in the appropriate class interval for each observation. The result is a histogram of X's instead of bars. This is used only for actual frequencies. It is often drawn transposed with the class intervals on the vertical axis, thus appearing somewhat like a stem and leaf graph.

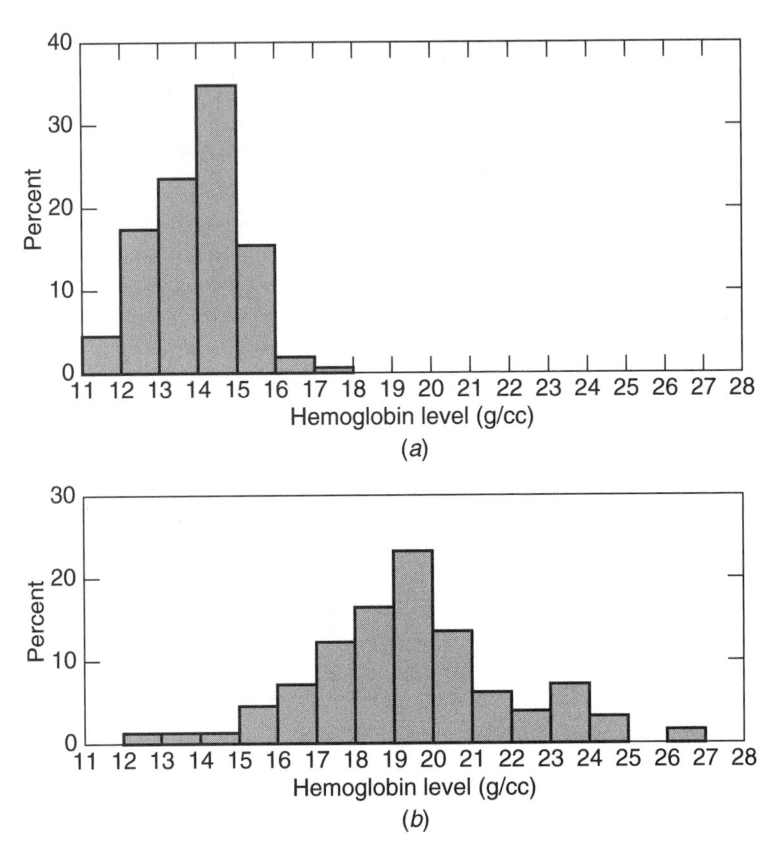

Fig. 3.1. Distributions of hemoglobin levels of mine workers. (*a*) Low-altitude mine workers. (*b*) High-altitude mine workers.

3.2.2 The Histogram: Unequal Class Intervals

When drawing a histogram from a set of data with unequal class intervals, we must first adjust for the length of the class intervals in order to avoid a graph that gives a misleading impression. Table 3.6 shows the age distribution of 302 hospital deaths from scarlet fever in 1905–1914. Both the frequencies and proportions are given.

If a histogram had been plotted directly from the frequencies in the table, it would leave the impression that deaths from scarlet fever rise sharply from 9 to 10 years of age. This erroneous impression arises simply from the fact that there is a change in the length of the class interval. The eye seems naturally to compare the various areas on the graph, rather than just their heights. The number of deaths from 10 up to 15 years is 14. Fourteen deaths is larger than the number of deaths at either age 8 or 9. If we plotted a bar 14 units high for the number of deaths from 10 up to 15 years, we would make it appear that the number of deaths were greater among this age group than among either 8 and 9 year olds.

Table 3.6. Frequency Table of Ages of 302 Deaths from Scarlet Fever[a]

Age (years)	Number of Deaths	Relative Frequency
0–	18	0.06
1–	43	0.14
2–	50	0.17
3–	60	0.20
4–	36	0.12
5–	22	0.07
7–	21	0.06
8–	6	0.02
9–	5	0.02
10–	14	0.05
15–20	3	0.01
Sum	302	1.00

[a] Data for this table are extracted with permission from A. Bradford Hill, *Principles of Medical Statistics*, Oxford Press, New York, 1961, p. 52.

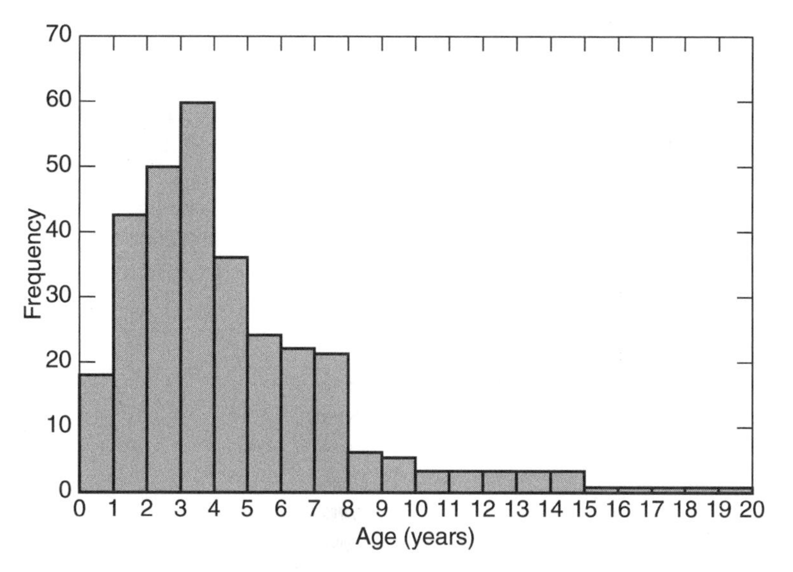

Fig. 3.2. Age of 302 deaths from scarlet fever: a histogram.

To correct this, we plot the number of deaths per year of the interval, rather than just the number of deaths. Since there are 14 deaths in the interval extending from 10 up to 15 years of age, there were 2.8 deaths per year of the interval. A single bar 2.8 units high is plotted above the entire class interval. The area of this bar is now 14 square units, as it should be. Frequencies have been plotted in Figure 3.2.

3.2.3 Areas under the Histogram

As mentioned earlier, the eye tends to compare the areas in a graph rather than the heights. In interpreting the histogram in Figure 3.1(a), for low-altitude workers, we might notice that the first, second, and third bars on the left side of the histogram seem to be somewhat less than one-half of the area of the entire histogram. From this, we conclude that less than one-half of the low-altitude hemoglobin levels were under 14 g/cc. In contrast, we see that only a very small percentage of hemoglobin levels in the high-altitude group was under 14 g/cc, since the area of the bars to the left of 14 is very small.

3.2.4 The Frequency Polygon

Instead of a histogram a frequency polygon is often made from a frequency distribution. It is made in the same way, except that instead of a bar of the proper height over each class interval, a dot is put at the same height over the midpoint of the class interval. The dots are connected by straight lines to form the frequency polygon. Sometimes the polygon is left open at each end, but usually it is closed by drawing a straight line from each end dot down to the horizontal axis. The points on the horizontal axis that are chosen to close the frequency polygon are the midpoints of the first class interval (on either end of the distribution) that has a zero frequency. Figure 3.3 gives a frequency polygon corresponding to the histogram of Figure 3.2.

The frequency polygon differs little from the histogram. It has the advantage of being somewhat quicker to construct by hand than the histogram. Further, to compare two frequency polygons at a glance may be easier than to compare two histograms. On the other hand, the frequency polygon has the disadvantage that the picture it gives of areas is somewhat distorted. The area over a certain class interval is no longer exactly

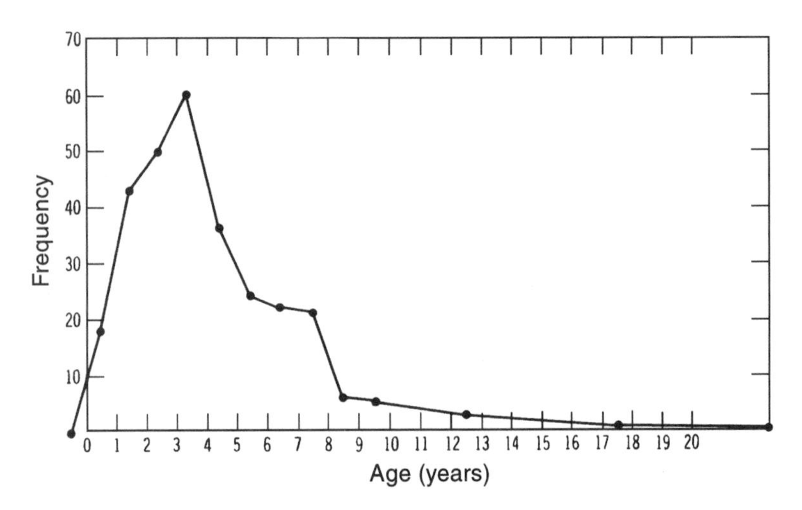

Fig. 3.3. Age of 302 deaths from scarlet fever: a frequency polygon.

equal to the proportion of the observations within that interval. The difference is slight, however, unless the number of class intervals is small.

Most computer programs do not provide frequency polygon plots but they can be often constructed as "line graphs" in spreadsheet software or statistical packages.

3.2.5 Discrete and Continuous Scales: Effect on Class Intervals

Some variables are thought of as rising from a *continuous* scale, whereas others arise from a discrete scale. Attributes such as height, weight, temperature, cholesterol level, age, and so on, are continuous; no matter how close we pick two possible heights, say 66.2 and 66.3 in., we can imagine a person whose height is between the two numbers.

On the other hand, if we consider the number of children per family, then for each family the count is either 0, 1, 2, or 3, and so on, and it is impossible to find a family with 2.4 children. The number of children is recorded on a *discrete* scale. If we make class intervals for discrete variables such as number of children, then we would like the center of each class interval to be a number such as 0, 1, 2, and so on, not 2.4. It is important when graphing discrete data to take into account what values are possible. It is also important to consider the purpose of the histogram. For example, in most cases it would not make sense to choose a class interval that included both 0 and 1 child since having 0 or 1 child is quite a different situation for a typical family. Other methods of classifying variables will be given in Section 4.5.3.

Hemoglobin level (Table 3.1) is a continuous variable. Nevertheless, it should be noted that even though it is continuous, the measurements were made only to the nearest tenth. In the recorded data, there can be no number between 19.2 and 19.3. Thus, as recorded, the data are always on a discrete scale. However, the underlying level is continuous irrespective of the precision of the measurement, so they can be regarded as essentially continuous.

3.2.6 Histograms with Small Class Intervals

If we imagine an immense set of data measured on a continuous scale, we may imagine what would happen if we picked a very small class interval and proceeded to make a frequency distribution and draw a histogram adjusting the vertical scale so that the area of all the bars totals 100%. With the bars of the histogram very narrow, the tops of the bars would get very close to a smooth curve, like that illustrated in Figure 3.4. If the large set of data that we are imagining is the population being studied, then the smooth curve will be called the frequency distribution or *distribution curve* of the population. A small or moderate-sized sample is not large enough to have its histogram well approximated by a smooth curve, so that the frequency distribution of such a sample is better represented by a histogram.

3.2.7 Distribution Curves

Distribution curves may differ widely in shape from one population to another. A few possibilities are shown in Figure 3.5. Such distribution curves have several properties

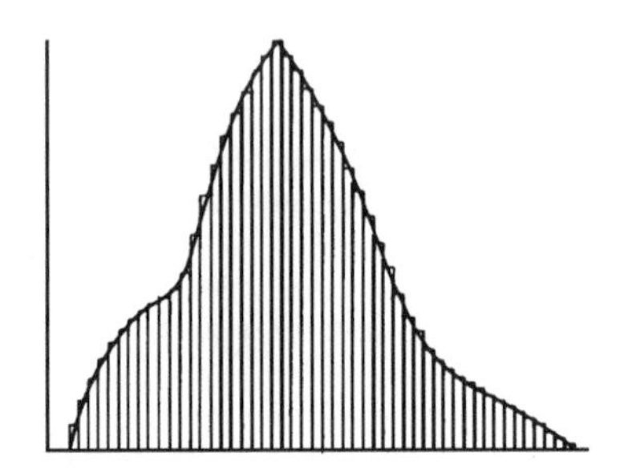

Fig. 3.4. Histogram with small class intervals.

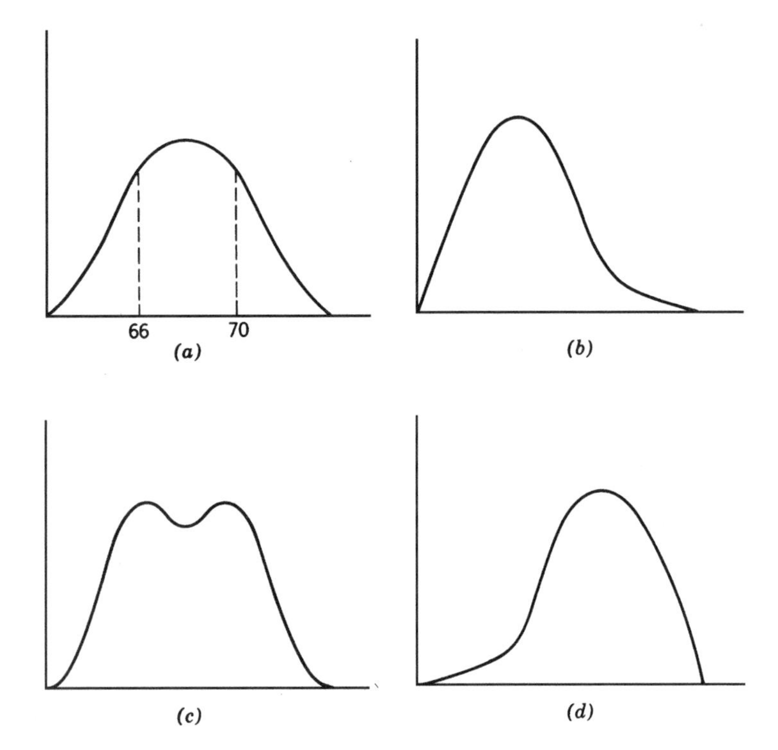

Fig. 3.5. Distribution curves.

in common: (1) The area under each one of them is equal to 100% or 1. (2) We may look at areas between the curves and the horizontal axis and interpret them as proportions of the individuals in the population. For example, if (a) is the frequency distribution for heights of a population, and if we judge that 20% of the area is above the portion of the horizontal axis to the left of 66 in., we decide that about 20% of the heights of the population are < 66 inches. Similarly, if the area from 66 to 70 in. is 60% of the entire area, then 60% of the heights are between 66 and 70 in.

The distribution curves depicted in Figure 3.5 are often described verbally. For example, (*a*) is called a symmetric distribution, where the right side is a mirror image of the left side; (*b*) has what is called a long tail to the right, sometimes called skewed to the right, and is common in biomedical and other data; (*c*) is called bimodal since it has two distinct high values with a dip between them; and (*d*) has a long tail to the left or is called skewed to the left. Forms (*c*) and (*d*) are less common in practice.

The frequency distribution of the population is an important concept. Actually, we rarely know the exact form of the distribution of the population we wish to study. We have a sample from the population, and we plot a histogram (or frequency polygon) from the sample; this can be regarded as giving some idea of the form of the frequency distribution of the population. One important use of the histogram or stem and leaf graph is to check visually if our sample has a form that is similar to what we expected in the population.

There are numerous other methods of depicting data. Texts on this subject include Cleveland [1985], Cleveland [1993], and Chambers et al. [1983]. For a discussion more oriented to the visual impact of the graphics, see Tufte [1990].

PROBLEMS

3.1 Make a stem and leaf graph of the first five rows of the blood cholesterol data in Table 2.1 (50 observations). Do you see an excess of 0's or 5's?

3.2 Use the stem and leaf graph from Problem 3.1 to assist you in ordering the data. From the ordered data, make a relative frequency table such as given in Table 3.5.

3.3 Make a histogram of the percentages from Problem 3.2. Does the distribution appear to be symmetric? Approximately what percentage of the cholesterol levels lie below 240 mg/100 mL? What percent lie above 250 mg/100 mL? Does a cholesterol level of 280 appear to be extremely high for this sample?

3.4 Draw a frequency polygon from Problem 3.2.

3.5 Using the results from Problem 3.2, draw the cumulative frequency in percent. What percent of the observations are below 240 mg/mL using this display?

3.6 Using an available statistical package, try to repeat what you obtained in Problems 3.1–3.5. Describe any problems you may have had.

3.7 Measure the length of the index finger in centimeters for each person in the class. With this data repeat Problem 3.1, 3.2, and answer the first two questions of Problem 3.3.

3.8 Make a histogram from a sample of 50 observations from Table A.1. Sketch what you would expect the histogram to look like if a very large sample was drawn.

3.9 From the following data on deaths from leukemia, compute the relative frequencies and draw a histogram displaying them.

Deaths from Childhood Leukemia by Age, United States, 1970[a]

Age (years)	Number of Deaths
< 1	68
1	82
2	98
3	137
4	169
5–9	684
10–14	434
Total	1672

[a] D.L. Levin, S.S. Devesa, J.D. Godwin II, and D.T. Silverman. 1974. *Cancer Rates and Risks*, 2nd ed. Washington, DC: US Department of Health, Education and Welfare, National Institutes of Health.

REFERENCES

Chambers, J. M., Cleveland, W. S., Kleiner, B., and Tukey, P. A. [1983]. *Graphical Methods for Data Analysis,* Belmont, CA: Wadsworth Incorporated, 7–69.

Cleveland, W. S. [1985]. *The Elements of Graphing Data*, Monterey, CA: Wadsworth Incorporated, 123–153.

Cleveland, W. S. [1993]. *Visualizing Data,* Summit, NJ: Hobart Press, 16–86.

Hill, B. A. [1961]. *Principles of Medical Statistics*, New York: Oxford Press, 52.

Tufte, E. R. [1990]. *Envisioning Information*, Cheshire, CT: Graphics Press.

CHAPTER FOUR

Measures of Location and Variability

In this chapter, we present the commonly used numbers that help describe a population or a sample of observations. Such numbers are called *parameters* if they describe populations; they are called *statistics* if they describe a sample.

The most useful single number or statistic for describing a set of observations is one that describes the center or the location of the distribution of the observations. Section 4.1 presents the most commonly used measures of location.

The second most useful number or statistic for describing a set of observations is one that gives the variability or dispersion of the distribution of observations. Several measures of variability will be defined in Section 4.2.

In Section 4.3, we give an example of the computation of the most widely used measures of location and variability from the frequency tables discussed in Chapter 3. We discuss the relationship between the results from the sample and the population for the mean and variance in Section 4.4. In Section 4.5, we discuss the reasoning behind the choice of the statistic to use for describing the center and variability of the distribution depending on what statements the investigator wishes to make and the characteristics of the variables in the data set. Section 4.6 presents two methods for displaying the sample statistics graphically.

4.1 MEASURES OF LOCATION

The number most often used to describe the center of a distribution is called the average or arithmetic mean. Here, we call it the mean to avoid confusion: there are many types of averages. Additional measures of location are described that are useful in particular circumstances.

4.1.1 The Arithmetic Mean

The Greek letter μ is used to denote the mean of a population; $\overline{X}$ (X-bar) is used to denote the mean of a sample. In general, Greek letters denote parameters of populations and Roman letters are used for sample statistics.

The mean for a sample is defined as the sum of all the observations divided by the number of observations. In symbols, if n is the number of observations in a sample, and the first, second, third, and so on, observations are called X_1, X_2, X_3,..., X_n, then $\overline{X} = (X_1 + X_2 + X_3 + \cdots + X_n)/n$.

As an example, consider the sample consisting of the nine observations 8, 1, 2, 9, 3, 2, 8, 1, 2. Here n, the sample size, is 9; X_1, the first observation, is 8; X_2, the second observation, is 1. Similarly, $X_3 = 2$, $X_4 = 9$, and so on, with $X_n = 2$. Then,

$$\begin{aligned}\overline{X} &= (8 + 1 + 2 + 9 + 3 + 2 + 8 + 1 + 2)/9 \\ &= 36/9 = 4\end{aligned}$$

The formula $\overline{X} = (X_1 + X_2 + \cdots + X_n)/n$ may be stated more concisely by using summation notation. In this notation, the formula is written $\overline{X} = \sum_{i=1}^{n} X_i/n$. The symbol $\sum$ means summation and $\sum_{i=1}^{n} X_i$ may be read as "the sum of the X_i's from X_1 to X_n", where n is the sample size. The formula is sometimes simplified by not including the subscript i and writing $\overline{X} = \sum X/n$. Here X stands for any observation, and $\sum X$ means "the sum of all the observations".

A similar formula $\mu = \sum X/N$ holds for μ, the population mean, where N stands for the number of observations in the population, or the population size. We seldom calculate μ for we usually do not have the data from the entire population. Wishing to know μ, we compute $\overline{X}$ and use $\overline{X}$ as an approximation to μ.

The mean has several interesting properties. Here, we mention several; others are given in Section 5.3. If we were to physically construct a histogram making the bars out of some material such as metal and not including the axes, the mean is the point along the bottom of the histogram where the histogram would balance on a razor edge.

The total sum of the deviations around the mean will always be zero. Around any other value the sum of the differences will not be 0 (see Weisberg [1992]). That is, $\sum(X - \overline{X}) = 0$. Further, the sum of the squared deviations around the mean is smaller than the sum of the squared deviations around any other value. That is, the numerical value of $\sum(X - \overline{X})^2$ will be smaller than if $\overline{X}$ were replaced by any other number.

Since the mean is the total divided by the sample size, we can easily obtain the mean from the total and vice versa. For example, if we weigh three oranges whose mean weight is 6 oz, then the total weight is 18 oz, a simple direct calculation. Other measures of location do not have this property.

When we change the scale that we use to report our measurements, the mean changes in a predictable fashion. For instance, if we add a constant to each observation, the mean is increased by the same constant. If we multiply each observation by a constant, as for example, in converting from meters to centimeters, then the mean in centimeters is 100 times the mean in meters.

In addition, the mean is easy to compute with computer programs or hand calculators. It is by far the most commonly used measure of location of the center of a distribution.

4.1.2 The Median

Another measure of location that is often used is the *median*. The median is the number that divides the total number of ordered observations in half. If we first order or rank the data from smallest to largest, then if the sample size is an odd number, the median is the middle observation of the ordered data. If the sample size is an even number the median is the mean of the middle two numbers. To find the median of the same data set of observations used in Section 4.1.1 (8, 1, 2, 9, 3, 2, 8, 1, 2) we order them to obtain 1, 1, 2, 2, 2, 3, 8, 8, 9. With nine observations, n is an odd number and the median is the middle or fifth observation. It has a value of 2. If n is odd, the median is the numerical value of the $(n + 1)/2$ ordered observation. If the first number were not present in this set of data, then the sample size would be even and the ordered observations are 1, 2, 2, 2, 3, 8, 8, 9; then the median is the mean of the fourth and fifth number observations or $(2 + 3)/2 = 2.5$. In general, for n even, the formula for the median is the mean of the $n/2$ and $(n/2) + 1$ observations.

The same definitions hold for the median of a finite population, with n replaced by N. When the size of the population is so large that it must be considered infinite, another definition must be used for the median. For continuous distributions, it can be defined as the number below which 50% of the population lie. For example if we found the value of X that divided the area of a frequency distribution such that one-half of the area was to the right and one-half was to the left, we would have the median of an infinite population.

The median provides the numerical value of a variable for the middle or most typical case. If we wish to know the typical systolic blood pressure for patients entering a clinic, then the proper statistic is the median.

The median (m) has the property that the sum of the absolute deviations (deviations that are treated as positive no matter whether they are positive or negative) of each observation in a sample or population from the median is smaller than the sum of the absolute deviations from any other number. That is, $\sum |X_i - m|$ is a minimum. If an investigator wants to find a value that minimizes the sum of the absolute deviations, then the median is the best statistic to choose (see Weisberg [1992]).

The median can also be found in some instances where it is impossible to compute a mean. For example, suppose it is known that one of the laboratory instruments is inaccurate for very large values. As long as the investigator knows how many large values there are the median can be obtained since it is dependent on only the numerical value of the middle or middle two ordered observations.

4.1.3 Other Measures of Location

Other measures of the center of a distribution are sometimes used. One such measure is the *mode*. The mode is the value of the variable that occurs most frequently. In order to contrast the numerical values of the mean, mode, and median from a sample that has a distribution that is skewed to the right (long right tail), let us look at a sample of $n = 11$ observations. The numerical values of the observations are 1, 2, 2, 2, 2, 3, 3, 4, 5, 6, 7. The mode is 2, since that is the value that occurs most frequently. The median is 3, since that is the value of the sixth or $(n + 1)/2$ observation. The mean is 3.36. Though small, this sample illustrates a pattern common in samples that are skewed to the right. That is, the mean is greater than the median which in turn is greater than the mode.

Other means have been proposed that were developed especially to be used when it is suspected that some of the observations in the sample may be either incorrect or come from a different population. Means that are not sensitive to some "incorrect" data are computed. Most of these means give greater weight to observations that fall in the middle of a set of ordered observations and less weight to those at either end. The reason for less weight for extreme values (those farthest from the center of the distribution) is twofold. First, those observations have more numerical effect on the size of a mean and second, they may be incorrect data. Such observations are often called outliers or extreme values. One type of mean is called a trimmed mean. Here, symmetrically placed extreme values are omitted or trimmed off before the mean is computed. See Dixon and Massey [1983] for an example of their use. A comprehensive discussion of this topic is given in Hoaglin et al. [1983]; this reference requires a more extensive background in statistics.

4.2 MEASURES OF VARIABILITY

After obtaining a measure of the center of a distribution such as a mean, we next wonder about the variability of the distribution and look for some number that can be used to measure how spread out the data are. Two distributions could have the same mean and look quite different if one had all the values closely clustered about the mean and the other distribution was widely spread out. In many instances, the dispersion of the data is of as much interest as is the mean. For example, when medication is manufactured the patient expects that each pill not only contains the stated amounts of the ingredients on the average but that each pill contains very close to the stated amount. The patient does not want a lot of variation.

The concept of variation is a more difficult one to get used to than the center or location of the distribution. In general, one wishes a measure of variation to be large if many observations are far from the mean and to be small if they are close to the mean.

4.2.1 The Variance and the Standard Deviation

Starting with the idea of examining the deviation from the mean, $(X - \overline{X})$ for all values of X, we might first think of simply summing them. That sum has been shown to be zero. We could also try summing the absolute deviations from the mean, $\sum |X_i - \overline{X}|$. That is, convert all negative values of $(X - \overline{X})$ to positive numbers and sum all values.

This might seem like a promising measure of variability to
are not so easy to work with as squared deviations $(X - \overline{X})$
desirable properties of the squared differences. (Note that the sq
also always positive.)

The sample *variance* is defined as the sums of squares of the differences b
each observation in the sample and the sample mean divided by one less than t
number of observations. The reason for dividing by $n - 1$ instead of n is given later in Section 4.4, but for now it is sufficient to remark that $n - 1$ is part of the definition of the variance.

The variance of the sample is usually denoted by s^2 and the formula is written out as

$$s^2 = \frac{\sum(X - \overline{X})^2}{n - 1}$$

The square root of the variance is also used. It is called the *standard deviation*. The formula for the standard deviation is

$$s = \sqrt{\frac{\sum(X - \overline{X})^2}{n - 1}}$$

When computing the sample variance with some hand calculators, it is sometimes easier to use an alternative formula for the variance. The alternative formula can be given as

$$s^2 = \frac{\sum X^2 - n(\overline{X})^2}{n - 1}$$

Table 4.1 shows the calculations needed to obtain the variance and standard deviation for the small sample of nine observations used in Section 4.1.1 to illustrate the calculations for the mean. We know that the mean of this small sample is 4 and calculate that the variance is $s^2 = \sum(X - \overline{X})^2/(n - 1) = 88/8 = 11$. The standard deviation is the square root of the variance, or $s = 3.317$.

The variance of a population is denoted by σ^2 (sigma squared) and the standard deviation by σ (sigma). If we have the entire population, we use the formula $\sigma^2 = \sum(X - \mu)^2/N$, where N is the size of the population. Note that N is used rather than $N - 1$. In other words, we simply compute the average of the squared deviations around the population mean.

The mean is relatively easy to interpret since people often think in terms of averages or means. But the size of the standard deviation is initially more difficult to understand. We return to this subject later in Sections 4.4 and 5.2. For now, if two distributions have the same mean, then the one that has the larger standard deviation (and variance) is more spread out.

Variation is usually thought of as having two components. One is the natural variation in whatever is being measured. For example, with systolic blood pressure,

ble 4.1. Calculation of the Variance

X	$X - \overline{X}$	$(X - \overline{X})^2$
8	+4	16
1	−3	9
2	−2	4
9	+5	25
3	−1	1
2	−2	4
8	+4	16
1	−3	9
2	−2	4
$\sum X = 36$	$\sum (X - \overline{X}) = 0$	$\sum (X - \overline{X})^2 = 88$

we know that there is a wide variation in pressure from person to person, so we would not expect the standard deviation to be very small. This component is sometimes called the biological variation. The other component of the variation is measurement error. If we measure something inaccurately or with limited precision, then we may have a large variation due simply to measurement methods. For a small standard deviation, both of these contributors to variation must be small.

Changing the measurement scale of the sample data affects the standard deviation and the variance though not in the same way that the mean is changed. If we add a constant to each observation, the standard deviation does not change. All we have done is shift the entire distribution by the same amount. This does not change the standard deviation or the variance. If we multiply each observation by a constant, say 100, the size of the standard deviation will also be multiplied by 100. The variance will be multiplied by 100 squared since it is the square of the standard deviation.

It should also be noted that knowing the mean and standard deviation does not tell us everything there is to know about a distribution. The four figures given in Figure 3.5 may all have the same mean and standard deviation but nevertheless they look quite different. Examining numerical descriptions of a data set should always be accompanied by examining graphical descriptions.

4.2.2 Other Measures of Variability

Some useful measures of variation are based on starting with *ordered* values for the variable under study. The simplest of these to obtain is the *range* that is computed by taking the largest value minus the smallest value, that is, $X_n - X_1$. Unlike the standard deviation the range tends to increase as the sample size increases. The more observations you have, the greater the chance of very small or large values. The range is a useful descriptive measure of variability and it has been used in certain applications such as quality control work in factories. It has mainly been used where repeated small samples of the same size are taken. It is easy to compute for a small sample and if the sample sizes are equal, then the problem of interpretation is less. The range is commonly included in the descriptive statistics output of statistical programs so it is readily available.

In many articles, the authors include the smallest and largest value in the tabled output along with the mean since they want to let the reader know the limits of their measurements. Sometimes, unfortunately, they fail to report the standard deviation. The range can be used to gain a rough approximation to the standard deviation. If the sample size is very large, say > 500, then simply divide the range by 6. If the sample size is 100–499 observations, divide the range by 5; if it is 15–99 observations, divide the range by 4; if it is 8–14, divide the range by 3; if the sample size is 3–7, divide by 2; and if it is a range of 2 numbers divide the range by 1.1. (Note that more detailed estimates can be obtained from Table A-8b(1) in Dixon and Massey [1983]). Such estimates assume that the data follow a particular distribution called the normal distribution that will be discussed in Chapter 5.

If outliers are suspected in the data set, then the range is a poor measure of the spread since it is computed from the largest and smallest values in the sample. Here, an *outlier* is defined as an observation that differs appreciably from other observations in the sample. These outliers can be simply errors of measurement or of recording data, or they can be the results of obtaining an observation that is not from the same population as the other observations. In any case, using a measure of variation that is computed from largest and smallest values is risky unless outliers do not exist or have been removed before computing the range.

A safer procedure is to obtain a type of range that uses observations that are not the largest and smallest values. The *interquartile range* (IQR) is one that is commonly used. The interquartile range is defined as IQR$= Q_3 - Q_1$. Three quartiles divide the distribution into four equal parts with 25% of the distribution in each part. We have already introduced one of the quartiles, Q_2 which is the median. The quartile Q_1 divides the lower half of the distribution into halves; Q_3 divides the upper half of the distribution into halves. Quartiles are computed by first ordering the data and the location of Q_1 is $.25(n+1)$, Q_2 is $.50(n+1)$, and Q_3 is $.75(n+1)$. The interquartile range is available in many statistical programs. The Q_1 and Q_3 quartiles are not easy measures to compute by hand as they often require interpolation. (Interpolation is a method of estimating an unknown value by using its position among a series of known values. An example of interpolation will be given in Section 5.2.2.) Since the quartiles are not sensitive to the numerical values of extreme observations (the extreme observations are simply counted as being either above Q_3 or below Q_1), they are considered measures of location resistant to the effect of outliers.

Note that the numerical value of the difference between the median and Q_1 does not have to equal the difference between Q_3 and the median. If the distribution is skewed to the right, then usually Q_3 minus the median is larger than the median minus Q_1. But the *proportion* of observations is the same.

For small samples, fourths are simpler measures to compute. If n is even, we simply compute the median of the lower and upper half of the ordered observations and call them the lower fourth and the upper fourth, respectively. If n is odd, we consider the middle measurement or median to be part of the lower half of measurements and compute the median of these measurements or Q_1. Then assign the median to the upper half of the measurements and compute the median of the upper half or Q_3. Another name for fourths is hinges. The difference between the upper and lower fourth (called the fourth-spread), should be close but not necessarily equal to the interquartile range since the quartiles are not necessarily equal to the fourths. For a more complete discussion of fourths, see Hoaglin et al. [1983].

Quartiles or fourths are often used when the distribution is skewed or outliers are expected. Also, in Section 4.5 an additional reason for using them is given.

4.3 COMPUTING THE MEAN AND STANDARD DEVIATION FROM A FREQUENCY TABLE

We will use hemoglobin levels measured on 90 high-altitude miners to illustrate computing the mean and standard deviation from a frequency table. This method of calculation can be used in cases where an author includes a frequency table in an article but does not present the raw data. It is not recommended for computations if the raw data are available since it is simpler to enter the data into a statistical package program and ask for all available descriptive measures. We will assume that all the observations in each class interval can be represented by the midpoint of that class interval. The midpoint of an interval is the mean of the smallest and largest measurements that could be included in the interval. In this data set, the observations have been recorded to the nearest tenth of a unit so that for the first class interval 12.0 could represent any number between 11.95 and 12.05; similarly 13.9 could represent any value between 13.85 and 13.95. The midpoint of the interval is $(11.95 + 13.95)/2$ or 12.95. The midpoint is denoted by X.

To calculate the sum of the observations in the first interval, add the midpoint twice since there are two observations in that interval, 14.95 five times, and so on. This is easily done by making a column headed fX in which the products 12.95 times 2,

Table 4.2. Calculations of the Mean and Standard Deviation for Hemoglobin Levels of High-Altitude Miners from a Frequency Table

Class Interval (g/cc)	Midpoint X	f	fX	$X - \overline{X}$	$(X - \overline{X})^2$	$f(X - \overline{X})^2$
12.0–13.9	12.95	2	25.90	−6.333	40.107	80.214
14.0–15.9	14.95	5	74.75	−4.333	18.775	93.874
16.0–17.9	16.95	17	288.15	−2.333	5.443	92.529
18.0–19.9	18.95	36	682.20	−0.333	1.109	3.992
20.0–21.9	20.95	17	356.15	1.667	2.779	47.241
22.0–23.9	22.95	9	206.55	3.667	13.447	121.022
24.0–25.9	24.95	3	74.85	5.667	32.115	96.345
26.0–27.9	26.95	1	26.95	7.667	58.783	58.783
$\sum$		90	1735.50			594.000

14.95 times 5, and so on, are recorded. Then, $\overline{X}$ is the sum of fX column divided by n or $1735.50/90 = 19.28$ g/cc. Note that $\sum f = n$. See Table 4.2 for details.

For the variance, we first compute $(X - \overline{X})$ using the midpoint as X and 19.28 as $\overline{X}$ for each class interval. Next, we square each $(X - \overline{X})$. Again, as with the mean in summing $\sum(X-\overline{X})^2$, we must add each $(X-\overline{X})^2$ f times, thus necessitating a column $f(X - \overline{X})^2$. We then compute $s^2 = \sum f(X - \overline{X})^2/(n - 1) = 594/89 = 6.674$ to obtain the variance and take the square root to compute the standard deviation or $s = 2.583$. Alternatively, using a hand calculator, we can compute the variance as

$$s^2 = \frac{\sum fX^2 - n\overline{X}^2}{n - 1}$$

4.4 SAMPLING PROPERTIES OF THE MEAN AND VARIANCE

To illustrate the behavior of the means and variances of samples, a population consisting of just four numbers (2, 10, 4, 8) is considered. The population has a mean $\mu = 24/4 = 6$ and a variance of $\sum (X - \mu)^2/N = 40/4 = 10$. All possible samples of size 2 from this small population are given in the first column of Table 4.3. There are 16 possible samples since we have sampled with replacement. The means and variances have been calculated from each sample and are listed in the second and last column of the table. These columns are labeled $\overline{X}$ and s^2.

The 16 sample means may be considered to be a new population —a population of sample means for samples of size 2. The new population contains 16 numbers, and so has a mean and variance. The mean and variance of this new population of means will be denoted by $\mu_{\overline{X}}$ and $\sigma^2_{\overline{X}}$, respectively. The mean of the sample means is calculated by summing the means in the second column and dividing by 16 or the number of means. That is,

$$\mu_{\overline{X}} = 96/16 = 6$$

The variance of the sample means is computed from

$$\sigma^2_{\overline{X}} = \frac{\sum (\overline{X} - \mu_{\overline{X}})^2}{N} = 80/16 = 5$$

or by summing the next to the last column and dividing by $N = 16$.

The mean of the population is $\mu = 6$ and the mean of the sample means is $\mu_{\overline{X}} = 6$. It is no coincidence that $\mu_{\overline{X}} = \mu$; this example illustrates a *general principle*. If all possible samples of a certain size are drawn from any population and their sample means computed, then the mean of the population consisting of all the sample means is equal to the mean of the original population.

A *second general principle* is that $\sigma^2_{\overline{X}} = \sigma^2/n$; here $\sigma^2/n = 10/2 = 5$ and $\sigma^2_{\overline{X}} = 5$. The variance of a population of sample means is equal to the variance of the population of observations divided by the sample size.

Table 4.3. Sampling Properties of Means and Variances, an Illustration

Samples	$\overline{X}$	$\overline{X} - \mu_{\overline{X}}$	$(\overline{X} - \mu_{\overline{X}})^2$	s^2
2,2	2	−4	16	0
2,10	6	0	0	32
2,4	3	−3	9	2
2,8	5	−1	1	18
10,2	6	0	0	32
10,10	10	4	16	0
10,4	7	1	1	18
10,8	9	3	9	2
4,2	3	−3	9	2
4,10	7	1	1	18
4,4	4	−2	4	0
4,8	6	0	0	8
8,2	5	−1	1	18
8,10	9	3	9	2
8,4	6	0	0	8
8,8	8	2	4	0
Σ	96	0	80	160

It should be noted that 4 rather than 3 was used in the denominator in the calculation of σ^2. Similarly, 16 was used instead of 15 in calculating $\sigma^2_{\overline{X}}$. This is done when the variance is computed from the entire population.

The formula $\sigma^2_{\overline{X}} = \sigma^2/n$ is perfectly general if we take all possible samples of a fixed size and if we sample with replacement. In practice, we usually sample without replacement and the formula must be modified somewhat to be exactly correct (see Kalton [1983]). However, for samples that are a small fraction of the population, as is generally the case, the modification is slight and we use the formula $\sigma^2_{\overline{X}} = \sigma^2/n$ just as if we were sampling with replacement. In terms of the standard deviation, the formula becomes $\sigma_{\overline{X}} = \sigma/\sqrt{n}$.

Since the mean of all the $\overline{X}$ is equal to μ, the sample mean $\overline{X}$ is called an *unbiased* statistic for the parameter μ. This is simply a way of saying that $\mu_{\overline{X}} = \mu$.

In a research situation, we do not know the values of all the members of the population and we take just one sample. Also, we do not know whether our sample mean is greater than or less than the population mean, but we do know that if we keep sampling always with the same sample size, in the long run the mean of our sample means will equal the population mean.

By examining the mean of all 16 variances, we can see that the mean of all the sample variances equals the population variance. We can say that the sample variance s^2 is an unbiased estimate of the population variance, σ^2. This is the *third general principle*.

As can be seen by examining the various s^2's in Table 4.3, they are highly variable. Our estimate of the population variance may be too high or too low or close to correct, but in repeated sampling, if we keep taking random samples, the estimates average to σ^2. The reason that $n - 1$ is used in the denominator of s^2, instead of n, is that it is desirable to have an unbiased statistic. If we had used n, we would have a biased estimate. The mean of s^2 from all the samples would be smaller than the population variance.

4.5 CONSIDERATIONS IN SELECTING APPROPRIATE STATISTICS

Several statistics that can be used as measures of location and variability have been presented in this chapter. What statistic should be used? In Section 4.5.1, we consider such factors as the study objectives and the ease of comparison with other studies. In Section 4.5.2, we consider the quality of the data. In Section 4.5.3, the importance of matching the type of data to the statistics used is discussed in the context of Stevens' system of measurements.

4.5.1 Relating Statistics and Study Objectives

The primary consideration in choosing statistics is that they must fit the major objectives of the study. For example, if an investigator wishes to make statements about the average yearly expenditures for medical care, then the mean expenditure should be computed. But if the hope is to decide how much a typical family spends, then medians should be considered.

Investigators should take into account the statistics used by other investigators. If articles and books on a topic such as blood pressure include means and standard deviations, then serious consideration should be given to using these statistics in order to simplify comparisons for the reader. Similarly, in grouping the data into class intervals, if 5-mmHg intervals have been used by others, then it will be easier to compare histograms if the same intervals are used. One need not slavishly follow what others have done but some consideration should be given if it is desired that the results be compared.

The choice of statistics should be consistent internally in a report or article. For example, most investigators who report means also report standard deviations. Those who choose medians are more apt to also report quartiles.

4.5.2 Relating Statistics and Data Quality

How clean is the data set? If the data set is likely to include outliers that are not from the intended population (either blunders, errors, or extreme observations), then statistics that are resistant to these problems should be considered. In this chapter, we have discussed the use of medians and quartiles, and briefly mentioned trimmed means, three statistics that are recommended for dealing with questionable data sets. Note that the mean and particularly the standard deviation is sensitive to observations whose numerical value is distant from the mean of the remaining values.

If an observation is a known error or blunder, most investigators will remove the observation or attempt to retake it, but often it is difficult to detect errors unless they result in highly unusual values.

4.5.3 Relating Statistics to the Type of Data

Measurements can be classified by type and, then depending on the type, certain statistics are recommended. Discrete and continuous measurements were discussed in Chapter 3. Another commonly used system is that given by Stevens. In Stevens' system, measurements are classified as *nominal, ordinal, interval, or ratio* based on what transformations would not change their classification. Here, we will simply present the system and give the recommended graphical diagrams and statistics. We do not recommend that this be the sole basis for the choice of statistics or graphs but it is another factor to consider.

In Stevens' system, variables are called *nominal* if each observation belongs to one of several distinct categories that can be arranged in any order. The categories may or may not be numerical although numbers are usually used to represent them when the information is entered into the computer. For example, the type of maltreatment of children in United States is classified into neglect, physical abuse, sexual abuse, emotional abuse, medical neglect, and others. These types could be entered into a statistical package using a word for each type or simply could be coded 1, 2, 3, 4, 5, or 6. But note that there is no underlying order. We could code neglect as a 1 or a 2 or a 3, and so on, it makes no difference as long as we are consistent in what number we use.

If the categories have an underlying order (can be ranked), then the variable is said to be *ordinal*. An example would be classification of disease condition as none, mild, moderate, or severe. A common ordinal measure of health status is obtained by asking respondents if their health status is poor, fair, good, or excellent. When the information for this variable is entered into the computer, it will likely be coded 1, 2, 3, 4. The order of the numbers is important but we do not know if the difference between poor health and fair health is equivalent in magnitude to the difference between fair health and good health. With ordinal data we can determine whether one outcome is greater than or less than another but the magnitude of the difference is unknown.

An *interval* variable is not only ordered but has equal intervals between successive values. For example, temperature in degrees on a Fahrenheit or a Celsius scale has a fixed unit of measurement (degrees). The difference between 14° and 15° is the same as between 16° and 17° or 1°. Note that 0° Fahrenheit or Celsius does not agree with no heat. Interval variables do not have natural zero points.

If the variable not only has equal intervals but also has a natural zero point, then it is called a *ratio* variable. Variables such as height, weight, and density have true zero points. Height has a naturally defined zero point on a ruler. We can multiply height in inches by a constant, say 2.54, and get height in centimeters and still have a ratio variable. If we reported temperature using the Kelvin scale, then it would be a ratio variable.

Table 4.4. Recommended Graphs and Statistics According to the Stevens' System

Scale	Graph	Location	Variability
Nominal	Pie and bar graphs	Mode	$p(1-p)/n$
Ordinal	Box plots	Median	Quartiles/range
Interval	Histograms	Mean	Standard deviation
Ratio	Histograms	Mean	Coefficient of variation

Recommended statistics and graphs for nominal, ordinal, interval, and ratio data are given in Table 4.4. It is important to note that for the statistics, the descriptive method is appropriate to that type of variable listed in the first column and *to all below it*. For example, The median is suitable for ordinal data and also for interval and ratio data. We will discuss graphics and statistics for nominal data in Chapters 8 and 9. Box plots will be discussed in Section 4.6.

In Stevens system, the mean and standard deviation are not recommended for nominal or ordinal data. This clearly makes sense for nominal data where even if the data is coded with successive numbers, the ordering is artificial. For example, if we coded hospital A as 1, hospital B as 2, and hospital C as 3, then if we compute and report a mean and standard deviation for a variable called hospital it obviously lacks any rational meaning. Although Stevens has shown that problems exist in using means and standard deviations for ordinal data, in practice it is sometimes done. Stevens recommends medians, quartiles, fourths, or ranges for ordinal data.

The *coefficient of variation* is a measure that is often used with ratio data when authors want to describe how variable their data are. It is computed by dividing the standard deviation by the mean. The coefficient of variation provides a measure of variation that is corrected for the size of the mean. If the coefficient of variation was a large value such as 1.0, then most investigators would decide that the observations are highly variable. But if the coefficient of variation was < 0.10, then for most work this would be considered a small variation (although likely not small enough for quality control results). The coefficient of variation should not be used unless there is a natural zero point. Note that when there is no natural zero point sometimes a commonly recognized lower limit to the numerical values can be subtracted from each observation so that the resulting observations have an approximate natural zero point.

4.6 COMMON GRAPHICAL METHODS FOR DISPLAYING STATISTICS

Stem and leaf charts and histograms for displaying the observations were discussed in Chapter 3. Here, we will present displays of two other commonly used graphs, box plots (originally called box and whisker plots) and mean comparison charts.

Box plots were originally designed to display medians, fourths, and extreme values (see Tukey [1977]) but the various statistical programs do not all display the same quantities so users should check what is displayed in their programs. In the *box plot*

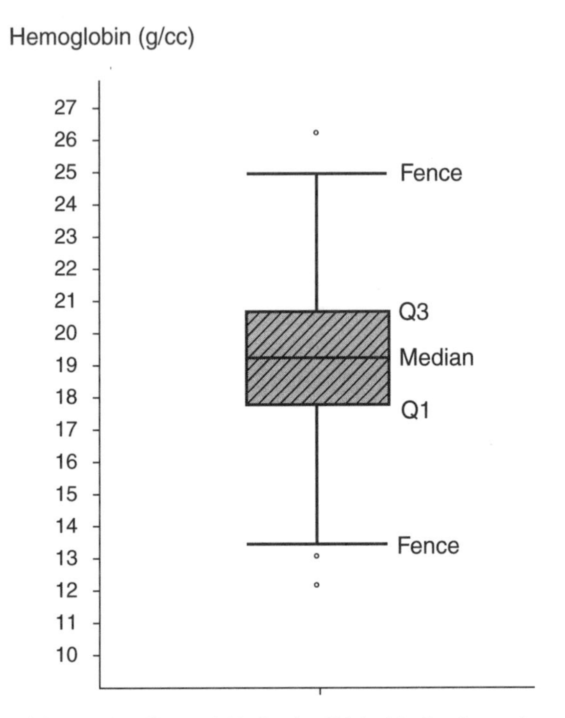

Fig. 4.1. Box plot of hemoglobin levels of high-altitude mine workers.

displayed in Figure 4.1, the horizontal line inside the box displays the median, the upper edge of the box is Q_3, and the lower edge of the box is Q_1. The length of the box is the IQR. The vertical lines above and below the box with the horizontal line at their end are the "whiskers". The horizontal lines are called fences. The upper fence is at ($Q_3 + 1.5$(IQR)) or the largest X, whichever is lower. The lower fence is at ($Q_1 - 1.5$(IQR)) or the smallest X, whichever is higher. Values that are outside the fences are considered possible extreme values or outliers. Here, we have depicted a box plot of the hemoglobin levels for mine workers at high elevations. Had the distribution of hemoglobin levels been skewed to the right, the portion of the box between the median and Q_3 would have been larger than the portion between the median and Q_1. There is one observation of 26.2 that is beyond the upper fence and two observations, 12.2 and 13.1, that are below the lower fence.

The *mean comparison* graphs depict the mean and the standard deviation of the observations or the standard deviation of the mean. In Figure 4.2, the mean is the small circle and the length of the bars represents the standard deviation for the high-altitude miner data. Note that the bars above and below the mean are the same length since both depict one standard deviation. This graph does not provide the information concerning the shape of the distribution that can be inferred from the box plot. Some computer packages also display a bar whose height indicates the mean with a whisker above it to display the standard deviation. Mean comparison graphs are often used

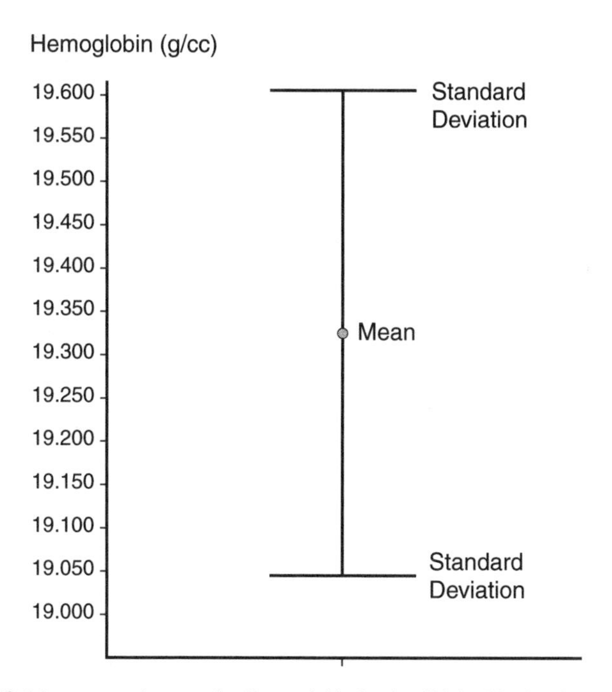

Fig. 4.2. Mean comparison graph of hemoglobin levels of high-altitude mine workers.

to compare several means side by side. For example, the means of the high-altitude miners could be displayed next to the low-altitude miners so that the reader could see the difference visually.

The graphical methods of comparison are especially useful for a presentation of data to an audience since most viewers can see the results in a graph more easily than in searching through a table.

PROBLEMS

4.1 The following data are ordered systolic blood pressures in millimeters of mercury (mmHg) from 48 young adult males:

87	106	114	120	129	140	155	183
93	107	116	122	133	141	155	194
101	107	117	122	133	146	162	197
104	109	118	125	134	146	167	204
105	110	118	125	135	148	173	212
105	114	119	128	138	152	176	230

(a) Compute the mean and standard deviation.
(b) Compute the median and quartiles.
(c) Compute the fourths.
(d) Display a histogram, a box plot, and a mean comparison graph of the above data.
(e) State whether or not you think the data follow a symmetric distribution and give the reason for your answer.
(f) Which observations might you check to see if they were outliers?

4.2 Compute the mean and standard deviation from Table 3.5 for the low-altitude miners.

4.3 For a population consisting of the numbers 1, 2, 3, 4, and 5, if all possible samples of size 4 were obtained, without performing calculations from the samples, what should the mean of all the sample means equal? What should the mean of the sample variances equal?

4.4 Compute the means and standard deviations for the three samples from Problem 2.1. Draw a mean comparison chart of your results. Place the means side by side on a single chart.

4.5 Combine all the means for Problem 4.4 from the class. Compute the variance of those means. From this variance, what would you estimate the variance of the data in Table 2.1 to be?

4.6 In a study of nursing home patients, the following data were obtained for each patient; age, gender, previous smoking status (smoker or non-smoker), reason for admission, nursing wing housed in, length of stay, and four point scale rating of health status. Classify each variable by whether it is nominal, ordinal, interval, or ratio according to Stevens' classification.

4.7 What is the difference between $\sum X^2$ and $(\sum X)^2$?

REFERENCES

Dixon, W. J. and Massey, F. J. [1983]. *Introduction to Statistical Analysis,* 4th ed., New York: McGraw-Hill, 380–381, 534.

Hoaglin, D. C., Mosteller, F., and Tukey, J. W. [1983]. *Understanding Robust and Exploratory Data Analysis,* New York: Wiley, 283–345, 35–36.

Kalton, G. [1983]. *Introduction to Survey Sampling,* Newbury Park, CA: Sage, 13.

Tukey, J. W. [1977]. *Exploratory Data Analysis,* Reading, MA: Addison-Wesley, 39–41.

Weisberg, H. F. [1992]. *Central Tendency and Variability,* Newbury Park, CA: Sage, 27–34, 22–27.

CHAPTER FIVE

The Normal Distribution

The statement that "men's heights are normally distributed" is meaningful to many who have never studied statistics. To some people, it conveys the notion that most heights are concentrated near a middle value, with fewer heights far away from this middle value; others might expect that a histogram for a large set of men's heights would be symmetric. Both of these ideas are correct.

More precisely, the statement means that if we take a very large simple random sample of men's heights, collect them into a frequency distribution, and draw a histogram of the relative frequencies, the histogram will be rather close to a curve that could have been plotted from a particular mathematical formula, the normal frequency function.[1]

In this chapter, we describe the normal distribution in Section 5.1 and in Section 5.2 show how to obtain areas under sections of the normal distribution using the table of the normal distribution. The major reasons why the normal distribution is used are given in Section 5.3. Three graphical methods for determining whether or not data are normally distributed are presented in Section 5.4. Finally, in Section 5.5 techniques are given for finding suitable transformations to use when the data are not normally distributed.

5.1 PROPERTIES OF THE NORMAL DISTRIBUTION

The mathematical formula for the normal frequency function need not concern us. It is sufficient to note some of the properties of the shape of this distribution and how we can use the area under the distribution curve to assist us in analysis.

[1]The formula is

$$Y = \frac{1}{\sqrt{2\pi}\sigma} e^{-(X-\mu)^2/2\sigma^2}$$

where (a) X is plotted on the horizontal axis, (b) Y is plotted on the vertical axis, (c) $\pi = 3.1416$, and (d) $e = 2.7183$.

First, we note that the area between a normal frequency function and the horizontal axis is equal to one square unit.

The curve is symmetric about the point at $X = \mu$ and is somewhat bell-shaped. It extends indefinitely far in both directions, approaching the horizontal axis very closely as it goes farther away from the center point. It is thus clear that men's heights could not possibly be exactly normally distributed, even if there were infinitely many heights, since heights below zero, for example, are impossible, and yet there is some area, however small, above the negative values of X. The most that we can expect is that a very large set of heights might be well approximated by a normal frequency function.

There are many normal curves rather than just one normal curve. For every different value of the mean and the standard deviation, there is a different normal curve. Men's heights, for example, might be approximately normally distributed with a mean height of 68 in., whereas the heights of 10-year-old boys could be normally distributed with a mean height of 60 in. If the dispersion of the two populations of heights is equal, then the two frequency functions will be identical in shape, with one merely moved 8 units to the right of the other (see Fig 5.1). It may very well be, however, that the variation among men's heights is somewhat larger than the variation among boys' heights. Suppose that the standard deviation for men's heights is 3 in., whereas the standard deviation for boys' heights is 2.5 in. Men's heights on the average will then be farther from 68 in. than boys' heights are from 60 in., so that the frequency curve for men will be flatter and more spread out than the other. Figure 5.2 shows two such normal frequency curves.

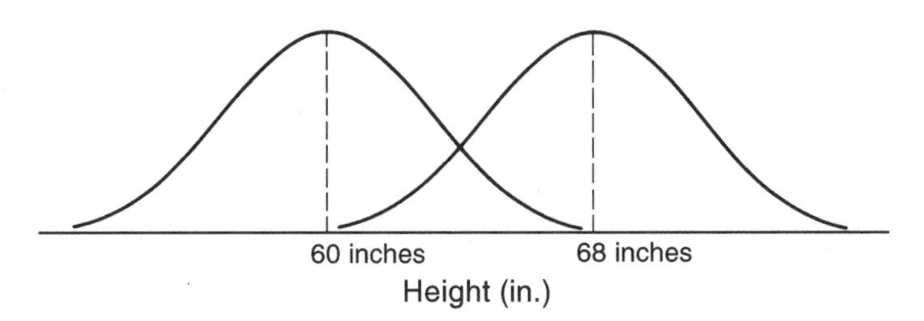

Fig. 5.1. Normal curves with different means.

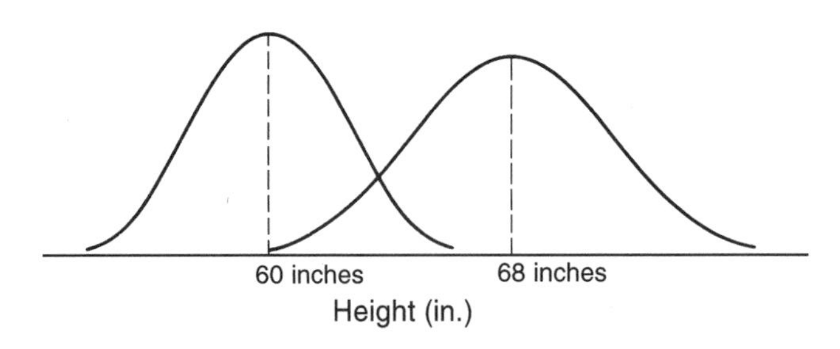

Fig. 5.2. Normal curves with different standard deviations.

For every pair of numbers μ and σ, then, a normal frequency curve can be plotted. Further, only one normal frequency curve can be plotted with these particular values for μ and σ. Thus if it is known that heights are normally distributed with a mean of 68 in. and a standard deviation of 3 in., the proper normal frequency distribution can be plotted and the entire distribution pictured. It was pointed out earlier that two sets of data might have the same means and the same standard deviations and still have frequency distributions very different from each other; obviously by knowing just the mean and standard deviation of a set of data, we could not draw a picture of the frequency distribution. However, if in addition to knowing the mean and standard deviation, we know that the data are *normally* distributed, then the entire frequency distribution can be drawn. For a population with a normal distribution, the mean and standard deviation tell the whole story.

5.2 AREAS UNDER THE NORMAL CURVE

As was discussed earlier in Section 3.2 on constructing histograms, the proportion of observations that fall between any two numbers, say X_1 and X_2, can be determined by measuring the area under the histogram between X_1 and X_2. Similarly, if a population is normally distributed, the proportion of X's that lie between X_1 and X_2 is equal to the area above the horizontal axis and under the normal curve and lying between X_1 and X_2.

It becomes important to measure these areas, and tables have been prepared that do so. Since it would be impossible to make a table for every normal curve, a transformation (a change in the mean and standard deviation) is made on the data so that a table for just one particular normal curve, which will be called the standard normal curve, will suffice.

5.2.1 Computing the Area under a Normal Curve

If the X's are normally distributed and their population mean is μ and their population standard deviation is σ, then if μ is subtracted from every observation, the resulting new population (the population of $X - \mu$) is also normally distributed; its mean is 0 and its standard deviation is σ. If, then, each $X - \mu$ is divided by σ, the $(X - \mu)/\sigma$ will again be normally distributed, with mean 0 and standard deviation 1. So, if we make the transformation $z = (X - \mu)/\sigma$, then the z's are normally distributed with mean 0 and standard deviation 1. The new variable z is called the standard normal variate, and the areas under its distribution curve are tabulated in Table A.2. In Table A.2, the columns are labeled $z[\lambda]$ and λ; λ percent of the area under the standard normal curve lies to the left of $z[\lambda]$. For example, under $z[\lambda]$ we read 2.00 and under λ in the same row we read .9772; thus 97.72% of the area lies below 2.00.

The use of Table A.2 is illustrated below, where we assume that the heights of a large group of adult men are approximately normally distributed; further, that the mean height is 68 in.; and that the standard deviation of the heights is 2.3 in.

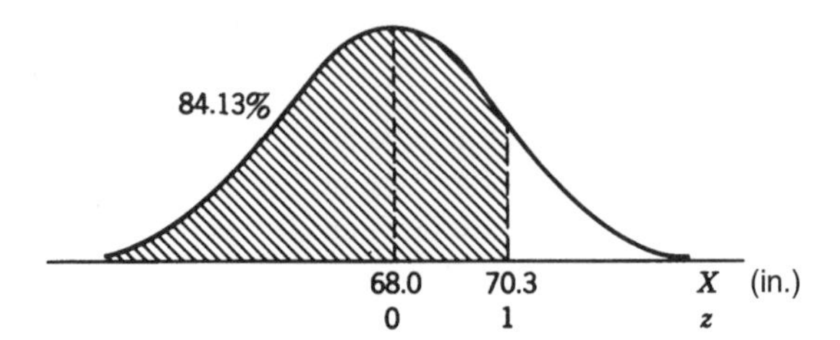

Fig. 5.3. Proportions of men's heights < 70.3 in.

1. What proportion of the men's heights is < 70.3 in.? Changing scale with the formula $z = (X - \mu)/\sigma$, we have

$$z = (70.3 - 68.0)/2.3 = 2.3/2.3 = 1$$

In Table A.2, we look for $z = 1$ under the column heading $z[\lambda]$; we then read across to the next column and find .8413. The column heading of this second column is λ. Since the areas under the curve are proportions, the proportion of men's heights < 70.3 in. is .8413. Or, equivalently, the percentage of men's heights below 70.3 in. is 84.13% (see Fig. 5.3).

2. The proportion of men's heights that are > 70.3 in. can be found by subtracting .8413 (the answer in **1**) from 1.0000 (the area under the entire curve). Thus 15.87% of the men are taller than 70.3 in. (see Fig. 5.4).

3. What proportion of the heights is < 65.7 in.? From

$$z = (X - \mu)/\sigma$$

we have $z = (65.7 - 68.0)/2.3 = -1$. Table A.2 does not give negative values of z, but since the curve is symmetric, the area below $z = -1$ is equal to the area above $z = 1$, or 15.87% (see Fig. 5.5.)

4. Subtracting both the percentage of heights that are < 65.7 in. and the percentage of heights that are > 70.3 in. from 100%, we obtain 68.26% as the percentage of heights lying between 65.7 and 70.3 in. Here, we see that approximately two-thirds of the observations lie between one standard deviation above the mean

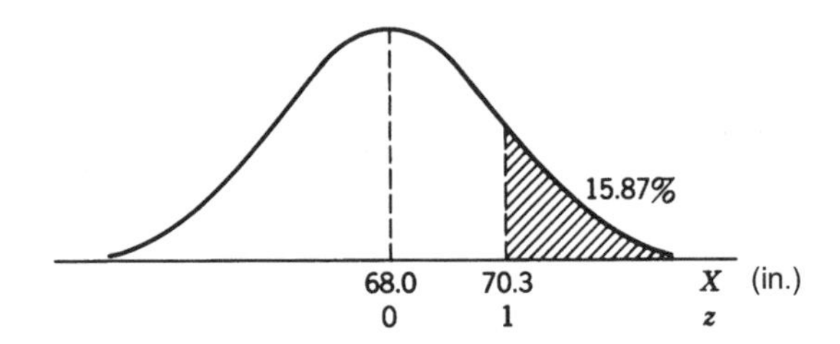

Fig. 5.4. Proportions of men's heights over 70.3 in.

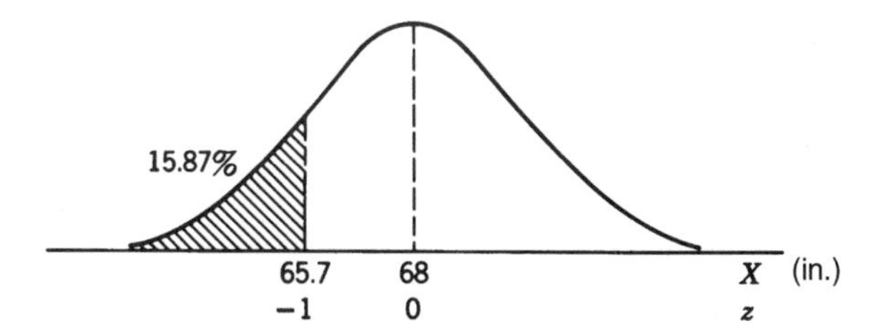

Fig. 5.5. Proportions of men's heights < 65.7 in.

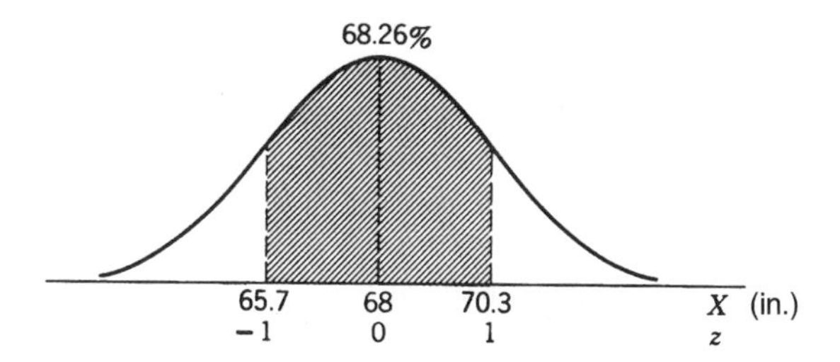

Fig. 5.6. Proportions of men's heights between 65.7 and 70.3 in.

and one standard deviation below the mean if the data are normally distributed (See Fig. 5.6).

5. It is sometimes of interest to know some percentiles for the distribution of men's heights. Suppose we wish to find the height that exceeds 99% of the men's heights, and is exceeded by just 1%. Such a number is called the *99th percentile*, or P_{99}. To find it, we look in the second column (i.e., the λ column) of Table A.2 and find .9901; in the first column we see 2.33, the approximate 99th percentile. Setting $\mu = 68.0$ and $\sigma = 2.3$ in $z = (X - \mu)/\sigma$, we can then solve for X, the height below which 99% of men's heights fall.

$$2.33 = \frac{X - 68.0}{2.3}$$

Multiplying both sides of this equation by 2.3,

$$5.359 = X - 68.0$$

Adding 68.0 to both sides of the equation,

$$X = 73.4 \text{ in.}$$

Thus 99% of men's heights are < 73.4 in.; that is, P_{99}, the 99th percentile, equals 73.4 in.

Many statistical programs print out certain percentiles such as the 1%, 2.5%, 5%, 10%, 25%, 50%, 75%, 90%, 95%, 97.5%, or 99% since these percentiles are considered to be useful in describing the distribution of the variables in the data set.

5.2.2 Linear Interpolation

Usually, when z is computed with the formula $z = (X - \mu)/\sigma$, it is found to be a number that is not given in Table A.2, but that lies between two values in the $z[\lambda]$ column of the table. We might reasonably proceed in one of four ways: First, use the closest z available if we are content with an answer that is not very accurate. Second, we could guess at a number between the two tabled values. Third, we could enter the value into a statistical program that reports z values. Note that many programs can give either an accurate z value for a given area or vice versa. Fourth, we can interpolate linearly to obtain a reasonably accurate result. Linear interpolation is more work but it can be done by hand.

We now illustrate the various methods of linear interpolation by finding the proportion of men's heights lying below 71 in. Here,

$$z = (71 - 68)/2.3 = 3/2.3 = 1.304$$

Note that 1.304 lies between the tabled values of $z[\lambda]$ of 1.30 and 1.31. From Table A.2, we write the information given in Table 5.1.

If not much accuracy is necessary, we might give as an answer .90 or 90%. Or for a little more accuracy, one could take a value of λ halfway between 0.9032 and 0.9049 since 1.304 is about one-half of the way between 1.30 and 1.31; we then choose 0.9040.

When more accuracy is desired, and one is working by hand, linear interpolation can be done as follows. We note that the distance between 1.310 and 1.300 is .010. Also, the distance between 1.304 and 1.300 is 0.004 or .4 of the distance between the tabled values (.004/.010 = .4).

The distance between .9049 and .9032 is .0017. Multiplying this distance by the same .4 results in .4(.0017) = .0007. Thus .0007 is .4 of the distance between .9032 and .9049. Finally, we add .0007 to .9032 to obtain .9039, an accurate estimate.

Usually, this computation is performed with the following formula:

$$\lambda = \lambda_1 + \frac{(\lambda_2 - \lambda_1)}{(z_2 - z_1)}(z - z_1)$$

Table 5.1. Numerical Example of Interpolation

	$z[\lambda]$	λ
Smaller tabled $z[\lambda]$ value	1.300	.9032
Computed z	1.304	?
Larger tabled $z[\lambda]$ value	1.310	.9049

Table 5.2. Symbols for Interpolation Formula

	$z[\lambda]$	λ
Smaller tabled $z[\lambda]$ value	$z_1 = 1.300$	$\lambda_1 = .9032$
Computed z	$z = 1.304$	$\lambda = ?$
Larger tabled $z[\lambda]$ value	$z_2 = 1.310$	$\lambda_2 = .9049$

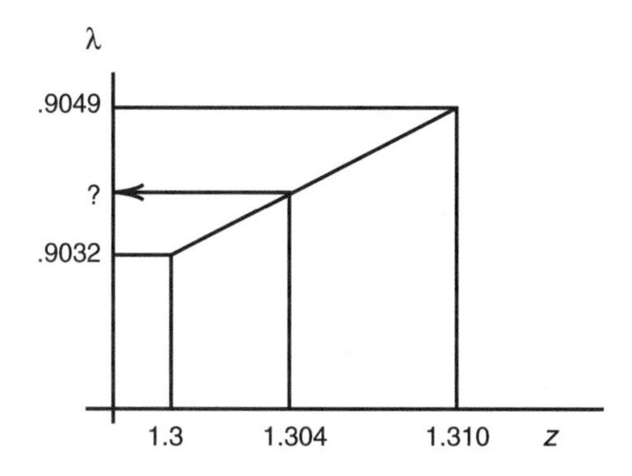

Fig. 5.7. Example of graphical interpolation.

Table 5.2 shows the relationship between the symbols in the formula for the interpolated λ and numbers in Table 5.1.

For men's heights, we have

$$\lambda = .9032 + \frac{(.9049 - .9032)}{(1.310 - 1.300)}(1.304 - 1.300) = .9039$$

The percentage of men's heights < 71 in. can be rounded off to 90.4%.

Figure 5.7 illustrates a simple graphical method of approximately the area λ corresponding to $z = 1.304$. When drawn on ordinary graph paper, λ is read from the vertical scale and can be used either as a check on the .9039 obtained from linear interpolation or as an approximation when not much accuracy is needed.

5.2.3 Interpreting Areas as Probabilities

In the foregoing example, areas under portions of the normal curve were interpreted as percentages of men whose heights fall within certain intervals. These areas may also be interpreted as probabilities. Where 84.13% of the area was below 70.3 in., the statement was made that 84.13% of men's heights are below 70.3 in. We may also say that if a man's height is picked at random from this population of men's heights, the probability is .8413 that his height will be < 70.3 in. Here the term probability is not defined rigorously. The statement, "the probability is .8413 that a man's height is < 70.3 in." means that if we keep picking a man's height at random from the population, time after time, the percentage of heights < 70.3 in. should come very close to 84.13%.

Similarly, we may say that the probability that a man's height is > 70.3 in. is .1587; the probability that a man's height is between 65.7 and 70.3 in. is .6826; the probability that a man's height is < 73.4 in. is .99; the probability that a man's height is > 73.4 in. is .01, and so on.

5.3 IMPORTANCE OF THE NORMAL DISTRIBUTION

One reason that the normal distribution is important is that many large sets of data are rather closely approximated by a normal curve. It is said then that the data are "normally distributed." We expect men's heights, women's weights, the blood pressure of young adults, and cholesterol measurements to be approximately normally distributed. When they are, the normal tables of areas are useful in studying them. It should be realized, however, that many large sets of data are far from being normally distributed. Age at death cannot be expected to be normally distributed, no matter how large the set of data. Similarly, data on income cannot be expected to be normally distributed.

Besides the fact that many sets of data are fairly well approximated by a normal distribution, the normal distribution is important for another reason.

For any population, if we choose a large sample size, draw all possible samples of that particular size from the population, and compute the mean for each sample, then *the sample means themselves are approximately normally distributed.* We use this a great deal in Chapter 6; at that time it will be clearer just why this makes the normal distribution so important. No matter how peculiar the distribution of the population, in general (under restrictions so mild that they need not concern us here), the means of samples are approximately normally distributed.

From Section 4.4, we know that the mean of a population of sample means is the same as the mean of the original population, whereas its standard deviation is equal to that of the original population standard deviation of the observations divided by $\sqrt{n}$, where n is the sample size. For n reasonably large, we know that the means are approximately normally distributed. Because a normal distribution is completely specified by its mean and its standard deviation, for large samples everything is known about the distribution of sample means provided μ and σ are known for the population. We need to know nothing about the shape of the population distribution of observations.

This remarkable fact will be useful in the succeeding chapters. At present, we can use it to answer the following question. If it is known that the mean cost of a medical procedure is \$5000 with a standard deviation of \$1000 (note that both the mean and standard deviation are considered population parameters), and we think of all possible samples of size 25, and their sample means, what proportion of these sample means will be between \$4600 and \$5400? Or, alternately, what is the probability that a sample mean lies between \$4600 and \$5400?

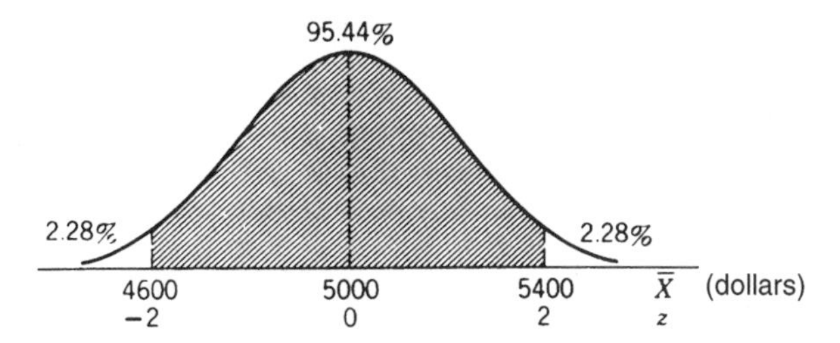

Fig. 5.8. Distribution of means from samples of size 25.

The population of $\overline{X}$'s from samples of size 25 is approximately normally distributed. From Chapter 4, the mean $\mu_{\overline{X}}$ is equal to \$5000 (since $\mu_{\overline{X}} = \mu$), and $\sigma_{\overline{X}}$ the standard deviation of the $\overline{X}$ distribution, equals $\$1000/\sqrt{25} = \$1000/5 = \$200$ (since $\sigma_{\overline{X}} = \sigma/\sqrt{n}$). Figure 5.8 shows the distribution for the population of $\overline{X}$, and the area of the shaded portion is equal to the proportion of the means between \$4600 and \$5400.

As usual, we make the transformation to z in order to be able to apply the normal tables. Now, however, we have $z = (\overline{X} - \mu_{\overline{X}})/\sigma_{\overline{X}}$, as the mean and standard deviation of the $\overline{X}$ distribution must be used. To find z at $\overline{X} = \$5400$, we have $z = (\$5400 - \$5000)/\$200 = 2$. From Table A.2, the area to the left of $z = 2$ is .9772. The area to the right of $z = 2$ must, by subtraction, equal .0228, and by symmetry, the area below $\overline{X} = \$4600$ is .0228. Subtracting the two areas from 1.0000, we obtain

$$1.0000 - 2(.0228) = 1.0000 - .0456 = .9544$$

Thus 95.44% of the sample means, for samples of size 25, lie between \$4600 and \$5400, and the probability that a sample mean is between \$4600 and \$5400 is .9544.

How large should a sample be to be called reasonably large? Twenty-five or larger is as satisfactory an answer as possible. The answer to this question must depend, however, on the answers to two other questions. First, how close to normality do we insist that the distribution of sample means be? Second, what is the shape of the distribution of the original population? If the original population is far from normal, the sample size should be larger than if it is from a population closer to normality. However, in assuming that samples of size 25 or greater have means that are normally distributed we make such a small error that in most work it can be disregarded.

However, if we do not know the population standard deviation and we desire to perform some of the analyzes described in Chapters 6, 7, 10, and 11, we will have to assume that the data are normally distributed.

5.4 EXAMINING DATA FOR NORMALITY

In this section, we present three graphical methods of determining if the variables in a data set are normally distributed. These graphical methods have the advantage of not only detecting nonnormal data but they also give us some insight on what to do to make the data closer to being normally distributed.

5.4.1 Using Histograms and Box Plots

One commonly used method for examining data to see if it is at least approximately normally distributed is to look at a histogram of the data. Distributions that appear to be markedly asymmetric or have extremely high tails are not normally distributed and the proportion of the area between any two points cannot be accurately estimated from the normal distribution. Here we will examine the data given in Problem 4.1, comparing it first with the normal distribution by means of a histogram.

In Figure 5.9, a histogram displaying the systolic blood pressures of 48 younger adult males is presented. It can be noted that the distribution is skewed to the right (the right tail is longer than the left). The mean of the distribution is $\overline{X} = 137.3$ and $s = 32.4$. A normal distribution with that mean and standard deviation is displayed along with the histogram. Many statistical programs allow the user to superimpose a normal distribution on a histogram. From Figure 5.9, the investigator may see whether the normal distribution should be used to estimate areas between two points. The fit is obviously poor, for the histogram's bars extend much higher than the normal curve between 100 and 130 mmHg and do not reach the curve between 130 to 190 mmHg. In examining histograms, the investigator must expect some discrepancies. This is especially true for small samples where the plotted percents may not fit a normal curve well even if the data were sampled from a normal distribution.

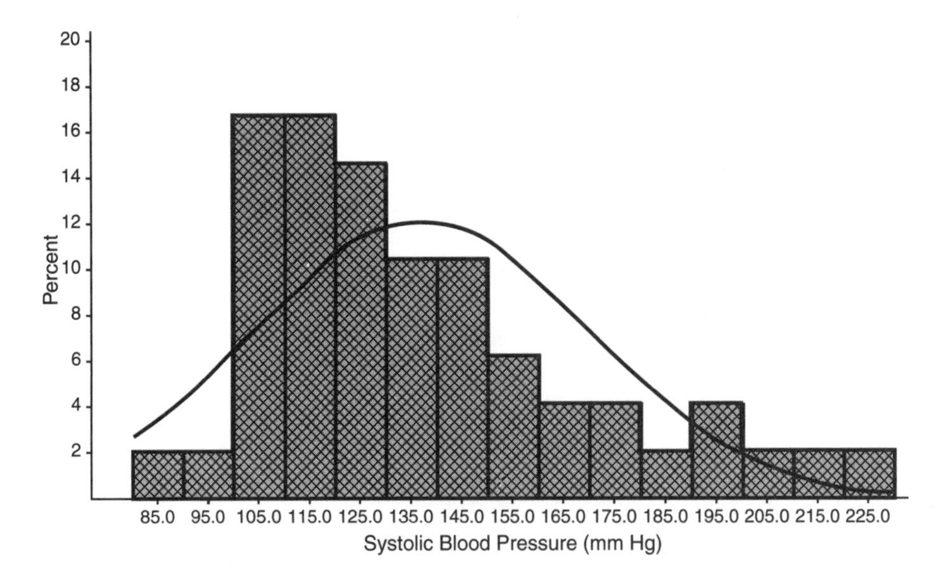

Fig. 5.9. Histogram of systolic blood pressures of adult males with normal curve.

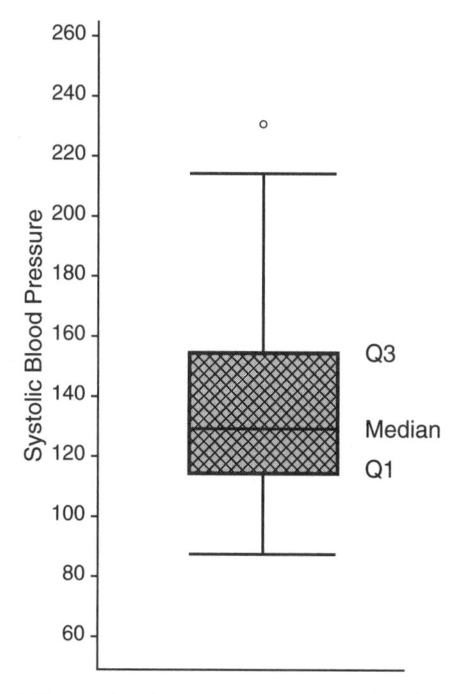

Fig. 5.10. Box plot of systolic blood pressures of adult males.

A box plot can be examined for the same data set (see Fig. 5.10). The distance between the median and Q_3 appears to be about twice the distance between the median and Q_1, a clear indication that the distribution is skewed to the right.

5.4.2 Using Normal Probability Plots

Another commonly used method to compare whether a particular variable is normally distributed is to examine the cumulative distribution of the data and compare it with that of a cumulative normal distribution. Table 3.5 illustrated the computation of the cumulative frequency in percent; in it, the percent cumulative frequency gave the percentage of miners with hemoglobin levels below the upper limit of each class interval. Rather than plot histogram the investigator can plot the upper limit of each class interval on the horizontal axis and the corresponding cumulative percent on the vertical axis.

Figure 5.11 is a cumulative plot of the normal distribution with $\mu = 0$ and $\sigma = 1$. At $\mu = 0$, the height of the curve is 50%. This height corresponds to the area below or to the left of the mean. (Note that since the normal distribution is symmetric that one-half of the area lies to the left of the population mean.) In general, the height of the cumulative distribution at any point along the X axis equals the area below or to the left of that point under the normal curve. The shape of the cumulative plot of a normal distribution is often described as being S shaped.

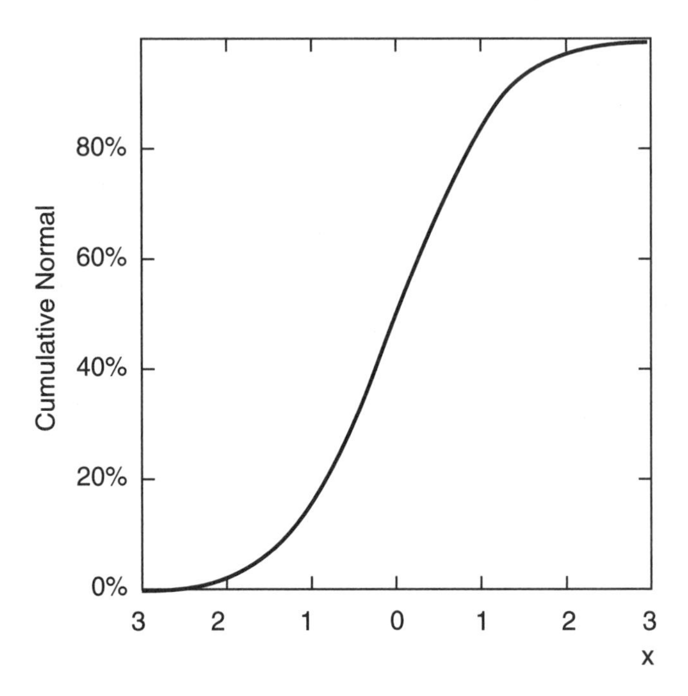

Fig. 5.11. Plot of cumulative normal distribution with mean zero and standard deviation of 1.

To make the comparison between the cumulative plot of the normal distribution and that of any other particular distribution easier to visualize, special graph paper has been made where the vertical scale has been nonuniformly altered so that the S shaped cumulative normal distribution curve of Figure 5.11 becomes a straight line. If we stretched the portion of the curve that lies above 50% progressively more, the higher it was above 50%, and the curve below 50% progressively more, the lower it is, we could obtain a straight line. The vertical scale is no longer an equal-interval scale. Special graph paper called *normal probability paper* is available; it has been scaled so that any normally distributed data plotted on it will be a straight line. Examples of such paper can be seen in Dixon and Massey [1983].

The simplest way to obtain normal probability plots is to use a statistical program. The plots available in these programs were programmed using different methods from that used with normal probability paper, but for visualizing if a variable is normally distributed they can be used in the same way. If the data are normally distributed, the plot will be a straight line. Such plots are usually called normal probability plots or normal quantile plots. Figure 5.12 shows a normal probability plot for the younger adult male systolic blood pressures from Problem 4.1. The variable X, systolic blood pressure, is shown on the horizontal axis. The expected values of X, given that the distribution follows a standard normal distribution, is on the vertical axis. If systolic blood pressure were normally distributed, the resulting points would lie approximately on a straight line. In examining these plots, the user should concentrate on the middle

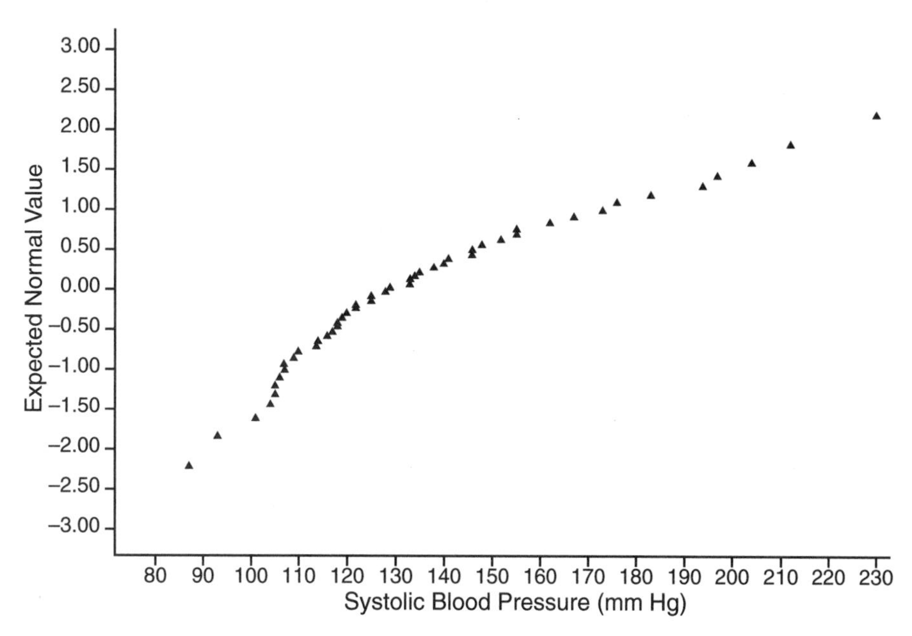

Fig. 5.12. Normal probability plot of systolic blood pressure of adult males.

90% and see if that portion is approximately a straight line. The middle points in Figure 5.12 resemble a curve that is lower at both ends (an upside down saucer).

The normal probability plots for data that are skewed to the right often show the curved pattern given in Figure 5.12. In contrast, when the normal probability plot is curved in the opposite direction, the distribution of the variable is likely to be skewed to the left. This latter pattern is less common in biomedical applications. If the variable has many extreme values (high tailed) the normal probability plot will look like an upside down saucer for low values of X and a right side up saucer for X greater than the mean (higher than expected for low values of X and lower than expected for high values of X). The shape of the curve is that of an inverted S.

5.5 TRANSFORMATIONS

A researcher with a variable that is not normally distributed has several options. First, the researcher may decide that the distribution of the observations is close enough to normality and proceed as if the data were normally distributed. Second, the researcher may use some statistical methods that do not assume a normal distribution (sometimes called nonparametric methods). A third option is to transform the data so that the resulting distribution is closer to a normal distribution. Transforming the data can sometimes produce an approximately normal distribution. In this process, the same transformation is applied to each observation of a given variable. Then, all analyses are performed on the transformed data and all results apply to the transformed data.

5.5.1 Finding a Suitable Transformation

Fortunately, the process of finding a suitable transformation to obtain approximately normal data is not a time consuming one using typical statistical programs. Some programs simply provide a list of transformations to choose from. The most commonly provided transformations for transforming data to normality are logarithmic and square-root transformations. Other statistical programs also allow the user to define their own transformation. When this option is not available, user defined transformations can be first performed using spreadsheet programs, and then the data can be transferred to a statistical program.

A commonly used transformation is taking the logarithm to the base 10 of each observation in the distribution that is skewed to the right. Note that the logarithm of a number X satisfies the relationship that $X = 10^Y$. Thus, the logarithm of X is the power Y to which 10 must be raised to produce X. The logarithm of X is usually abbreviated as $\log(X)$. The $\log(10)$ is 1 since $10 = 10^1$ and the $\log(100)$ is 2 since $100 = 10^2$. The log of 1 is 0 since $1 = 10^0$. Note that as X increases from 1 to 10, $\log(X)$ increases from 0 to 1, and if X goes from 10 to 100, $\log(X)$ goes from 1 to 2. For larger values of X, it takes an even greater increase in X for $\log(X)$ to increase much. For a distribution that is skewed to the right, taking the logarithm of X has the effect of reducing the length of the upper tail of the distribution and making it more nearly symmetrical.

The logarithm of any X that is ≤ 0 is undefined and the logarithm of any $X < 1$ but > 0 is negative. Note that $\leq$ signifies less than or equal to and $\geq$ signifies greater than or equal to. When X is < 1, a small positive constant A can be added first to X so that the logarithm of X plus a constant A [or $\log(X + \text{A})$] results in a positive number.

If the numbers being transformed are all quite large, then often a positive constant is subtracted first. This has the result of increasing the effect of taking the log transformation.

One general strategy for finding an appropriate transformation is the use of power transformations (see Tukey [1977] or Afifi and Clark [1996]). Consider the effects of taking X^P for various values of P. With $P = 2$, then large values of X become much larger. With $P = 1$, then there is no change in the value of $X (X^1 = X)$. With $P = 0.5$ we have the square-root transformation which has the effect of reducing the value of large values of X. Taking $P = 0$ results in the logarithm transformation (see Cleveland [1993] for an explanation), which has the effect of reducing large values of X also. With $P = -1$ we have $1/X$, which changes large values of X into small values and reverses the ordering of the data.

A general rule is that with a distribution that is skewed to the right as was systolic blood pressure, then we should try values of $P < 1$; the reduction in skewness to the right increases as P is decreased. That is, taking the logarithm of X reduces the amount of skewness to the right more than taking the square root transformation $\left(0 \text{ is } < \frac{1}{2}\right)$; taking the reciprocal of X or $P = -1$ reduces it even more.

With a distribution skewed to the left we should begin with a value of $P > 1$ and if necessary try larger values of P until the skewness is sufficiently reduced.

After each attempted transformation, a normal probability plot, histogram or stem and leaf plot, or a box plot should be examined.

5.5.2 Assessing the Need for a Transformation

An awkward questions is whether the data that appears to be nonnormal needs to be transformed. Most investigators do not use a transformation for slight departures from normality.

Several rules of thumb have been suggested to assist in deciding if a transformation will be useful. Two of these rules should only be used for ratio data. If the standard deviation divided by the mean is $< \frac{1}{4}$, it is considered less necessary to use a transformation. For example, for the systolic blood pressure data from Problem 4.1, the standard deviation is 32.4 and the mean is 137.3, so the coefficient of variation is $32.4/137.3 = .24$. This would be an indication that even if the points in Figure 5.12 do not follow a straight line, it is questionable if a transformation is needed. An alternative criteria is if the ratio of the largest to the smallest number is < 2, then a transformation may not be helpful. Here the ratio is $230/87 = 2.6$, so perhaps a transformation is helpful but it again seems borderline.

One reason for the reluctance to use a transformation is that for many readers, it makes the information harder to interpret. Also, users wishing to compare their information with that of another researcher who has not used a transformation, will find the comparison more difficult to make if they transform their data. For the same reason, researchers often try to use a transformation that has already been used on a particular type of data. For example, logarithms of doses are often used in biomedical studies.

Finding an appropriate transformation can also provide additional information concerning the data. When, for instance, a logarithmic transformation results in near normal data we know that our original data follows a skewed distribution called a lognormal distribution.

Often researchers find the best possible transformation for their data, and then perform their analyses both with and without the transformation used and see if the transformation affects the final results in any appreciable fashion. If it does not, the transformation may not be used in the final report.

PROBLEMS

5.1 Does the stem and leaf display of the hemoglobin data given in Table 3.2 appear to be approximately normally distributed?

Problems 5.2–5.7 refer to the following situation: A population of women's weights is approximately normally distributed with mean equal to 140 lb and standard deviation equal to 20 lb.

5.2 What percentage of women weigh between 120 and 160 lb?

5.3 If samples of size 16 are drawn from the population of women's weights, what percentage of the sample means lie between 120 and 160 lb?

5.4 What is the probability that a sample mean from a sample size 16 lies above 145 lb?

5.5 Find P_{95}, the 95th percentile for the population of women's weights.

5.6 Find P_{95}, the 95th percentile for the population of means for samples of size 16.

5.7 In Problems 5.4 and 5.6, what justification is there for using the tables of the normal distribution?

5.8 If the mean diastolic blood pressure in a certain age group is 80, and the standard deviation is 10, what is the probability that, in a sample of size 100, the sample mean is > 85?

5.9 What, if anything, do you need to know about the form of the distribution of diastolic blood pressures to use the normal tables in Problem 5.8?

5.10 If, from the population in Problem 5.8, you were to take a sample of size 100, and compute $\overline{X} = 85$, would you feel that some explanation is necessary? What possible explanations occur to you?

5.11 A quality control inspector examines some bottles of a herbal medication to determine if the tablets are all within stated limits of potency. The results were that 13 bottles had 0 tablets out of limits, 26 bottles had 1 tablet, 28 bottles had 2 tablets, 18 bottles had 3 tablets, 9 bottles had 4 tablets, 4 bottles had 5 tablets, and 2 bottles had 6 tablets out of limits. In all, $n = 100$ bottles were tested. Give two transformations that are suitable transformations to try that might achieve a distribution somewhat closer to a normal distribution for the number of tablets per bottle that are outside the stated limit.

REFERENCES

Afifi, A. A. and Clark, V. [1996]. *Computer-Aided Multivariate Analysis*, 3rd ed., London: Chapman & Hall, 48–64.

Cleveland, W. S. [1993]. *Visualizing Data,* Summit, NJ: Hobart Press, 56.

Dixon, W. J and Massey, F. J. [1983]. *Introduction to Statistical Analysis,* 4th ed., New York: McGraw-Hill Book Company, 64, 487.

Tukey, J. W. [1977]. *Exploratory Data Analysis,* Reading, MA: Addison-Wesley, 57–93.

CHAPTER SIX

Estimation of Population Means: Confidence Intervals

We consider first a very simple research situation in which we wish to estimate a population mean μ. Pediatricians have included a new substance in the diet of infants. They give the new diet to a sample of 16 infants and measure their gains in weight over a 1-month period. The arithmetic mean of the 16 gains in weight is $\overline{X} = 311.9$ g.

These specific 16 infant gains in weight observations are from a sample and are not the population that we are studying. That population consists of the infants from which the pediatricians took their sample in performing the study, and we wish to estimate μ, the population mean weight gain. The *target* population consists of similar infants who may receive the new diet in the future.

The *point estimate* (i.e., an estimate consisting of a single number) for the population mean μ is the sample mean, or 311.9 g. By now, however, enough has been discussed about sampling variation to make clear that μ is not exactly 311.9 g, and we wish to get some idea of what μ may reasonably be expected to be. To fill this need, we will compute a confidence interval for μ; the term, confidence interval, will be defined after the example is worked out.

In Section 6.1, we present an example of computing a confidence interval for a single mean when the population standard deviation, σ, is known. We also give a definition of confidence intervals and discuss the choice of the confidence level. In Section 6.2, the sample size needed to obtain a confidence interval of a specified length is given. We need to know σ in order to obtain a confidence interval using the normal distribution. Often σ is unknown. A distribution that can be used when σ is unknown is introduced in Section 6.3. This distribution is called the t distribution or the Student t distribution. The formula for the confidence interval for a single mean using the t distribution is presented in Section 6.4. Confidence intervals for the differences in two means when the data come from independent populations is given in Section 6.5. The case of paired data is discussed in Section 6.6.

Table 6.1. Weight Gain under Supplemented and Standard Diet

Supplemented Diet		Standard Diet	
Infant Number	Gain in Weight (g)	Infant Number	Gain in Weight (g)
1	448	1	232
2	229	2	200
3	316	3	184
4	105	4	75
5	516	5	265
6	496	6	125
7	130	7	193
8	242	8	373
9	470	9	211
10	195		
11	389		
12	97		
13	458		
14	347		
15	340		
16	212		
$n_1 = 16$	$\overline{X}_1 = 311.9$ $s_1^2 = 20{,}392$	$n_2 = 9$	$\overline{X}_2 = 206.4$ $s_2^2 = 7060$

6.1 CONFIDENCE INTERVALS

First, we illustrate how to compute a confidence interval for μ, the population mean gain in weight when we assume that σ is known.

6.1.1 An Example

To simplify the problem as much as possible, suppose that we have been studying the diet of infants for some time and that, from the large amount of data accumulated, have found the standard deviation of gains in weight over a 1-month period to be 120 g. It seems rather reasonable to believe that the standard deviation of the population under the new diet will be quite close to that under the old diet. We assume, therefore, that the population's standard deviation, under the new diet, is 120 g. The data for this example are given in Table 6.1 in the columns labeled Supplemented Diet.

Since the assumed population standard deviation is $\sigma = 120$ g, the standard deviation for the population of means of samples of size 16 is $\sigma_{\overline{X}} = \sigma/\sqrt{n} = 120/\sqrt{16} = 120/4 = 30$ g (see Section 4.4, where sampling properties of the mean and variance are discussed). In order to compute a 95% confidence interval for μ, we find the distance on each side of μ within which 95% of all the sample means lie. This is illustrated in Figure 6.1. From Table A.2, we find that 95% of all the z's lie between -1.96 and $+1.96$. The value of the sample mean $\overline{X}$ corresponding to $z = 1.96$ is obtained by solving $1.96 = (\overline{X} - \mu)/\sigma_{\overline{X}}$ for $\overline{X}$. This gives

$$\overline{X} = \mu + 1.96(30)$$

or

$$\overline{X} = \mu + 58.8$$

Similarly, corresponding to $z = -1.96$ we have $\overline{X} = \mu - 1.96(30)$. Thus 95% of all the $\overline{X}$'s lie within a distance of $1.96(30) = 58.8$ g from the mean μ. The 95% confidence interval for μ is

$$\overline{X} \pm 58.8$$

where $\pm$ signifies plus or minus. Substituting in the numerical value of the sample mean, we have

$$311.9 \pm 58.8$$

or

$$253.1 \text{ to } 370.7 \text{ g}$$

The measured sample mean $\overline{X}$, 311.9 g, may or may not be close to the population mean μ. If it is one of the $\overline{X}$'s falling under the shaded area in Figure 6.1, then the interval 311.9 ± 58.8 g includes the population μ. On the other hand, if 311.9 is one of the sample means lying farther away from μ in the unshaded area, then the confidence interval does not include μ. We do not know whether this particular $\overline{X}$ is close to μ or not, but we do know that in repeated sampling 95% of all the $\overline{X}$'s obtained will be close enough to μ so that the interval $\overline{X} \pm 58.8$ contains μ.

If we take a second sample of 16 infants a different $\overline{X}$ will be obtained and the 95% confidence interval will then be a different interval. The intervals obtained vary from one sample to the next. They are formed in such a way, however, that in the long run about 95% of the intervals contain μ.

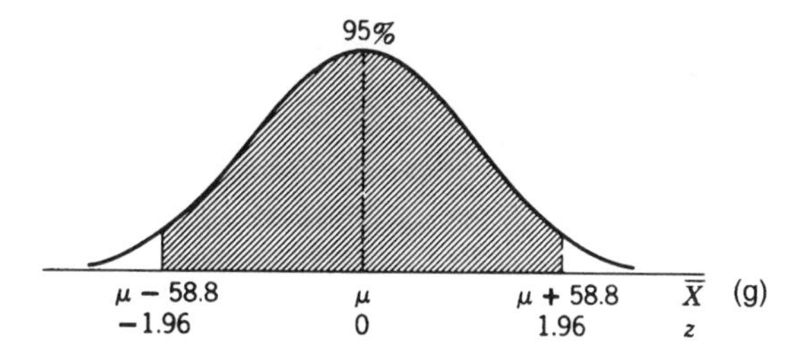

Fig. 6.1. Distribution of means from sample of size 16.

6.1.2 Definition of Confidence Interval

A 95% confidence interval for a population parameter is an interval obtained from a sample by some specified method such that, in repeated sampling, 95% of the intervals thus obtained include the value of the parameter. For further discussion of the meaning of confidence limits, see Dunn and Clark [1987], Rothman [1986], or Fisher and van Belle [1993].

The *confidence level* is 95%, and we say that we have 95% confidence that μ lies within the interval. For the interval 253.1–370.7 g, 253.1 g is the lower confidence limit, and 370.7 g is the upper confidence limit. The rule for obtaining a 95% confidence interval in this situation is

$$\overline{X} \pm 1.96\sigma_{\overline{X}}$$

or

$$\overline{X} \pm \frac{1.96\sigma}{\sqrt{n}}$$

Certain assumptions are made in obtaining this confidence interval. We assume that the 16 infants are a simple random sample from a population with a standard deviation equal to σ. Each infant measured is assumed to be independent of the other infants measured (for example, twins are excluded from the study). We also assume that $(\overline{X} - \mu)/\sigma_{\overline{X}}$ is normally distributed as is the case if X is normally distributed. With large samples or with a distribution close to the normal, we know that $\overline{X}$ is normally distributed and most investigators assume normality unless the sample is quite nonnormal in appearance.

6.1.3 Choice of Confidence Level

There was no necessary reason for choosing a 95% confidence interval; we might compute a 90% confidence interval or a 99% confidence interval.

For a 90% interval, we find from Table A.2 that 90% of the z's lie between ± 1.645, so that the 90% confidence interval is

$$311.9 \pm 1.645(30) = 311.9 \pm 49.4$$

or 262.5 to 361.3 g. In repeated sampling, about 90% of all such intervals obtained cover μ. A 99% confidence interval is

$$311.9 \pm 2.575(30) = 311.9 \pm 77.2$$

or 234.7–389.1 g.

If a higher confidence level is chosen, we have greater confidence that the interval contains μ; but on the other hand, we pay for this higher level of confidence by having a longer interval. The confidence levels generally used are 90, 95, and 99%.

6.2 NEEDED SAMPLE SIZE FOR A DESIRED CONFIDENCE INTERVAL

To obtain a short interval and at the same time to have one in which we have a high level of confidence, we must increase the sample size. In the simplified example under consideration, we can in the planning stages of the experiment calculate the length of a 95% confidence interval. Since the interval is $\overline{X} \pm 1.96\sigma/\sqrt{n}$, the length of the entire interval is $2(1.96)\sigma/\sqrt{n}$. If we wish this length to be only 60 g, we can solve the equation

$$60 = \frac{2(1.96)120}{\sqrt{n}}$$

$$\sqrt{n} = \frac{2(1.96)(120)}{60} = 7.84$$

$$n = 61.47 \quad \text{or} \quad 62$$

The required sample size is 62 since in calculating a sample size for a certain confidence level, one rounds up to a whole number rather than down.

In general, if we call the desired length of the interval L, the formula for n can be written as

$$n = \left(\frac{2z[\lambda]\sigma}{L}\right)^2$$

where $z[\lambda]$ is the tabled value from Table A.2 (1.645 for 90%, 1.96 for 95%, and 2.575 for a 99% confidence interval).

Calculations of this sort help us to determine in advance the size of sample needed. Estimating the necessary sample size is important; there is no "best" sample size applicable to all problems. Sample size depends on what is being estimated, on the population standard deviation σ, on the length of the confidence interval, and on the confidence level.

6.3 THE *t* DISTRIBUTION

In most research work, the value of the population variance or standard deviation is unknown, and must be estimated from the data. In the example in Section 6.1.1, we assumed we knew that σ equaled 120 g, and made use of the fact that the quantity $z = (\overline{X} - \mu)/(\sigma/\sqrt{n})$ has a standard normal distribution whose areas have been calculated and are available in Table A.2 or from a statistical program. In a more usual research situation, the population standard deviation σ is unknown and must be estimated by computing the sample standard deviation s. This requires the use of the quantity $t = (\overline{X} - \mu)/(s/\sqrt{n})$, which will be used instead of $z = (\overline{X} - \mu)/(\sigma/\sqrt{n})$.

The distribution of t (often called the Student's t distribution) does not quite have a normal distribution. The t distribution is bell-shaped and symmetric, as is the

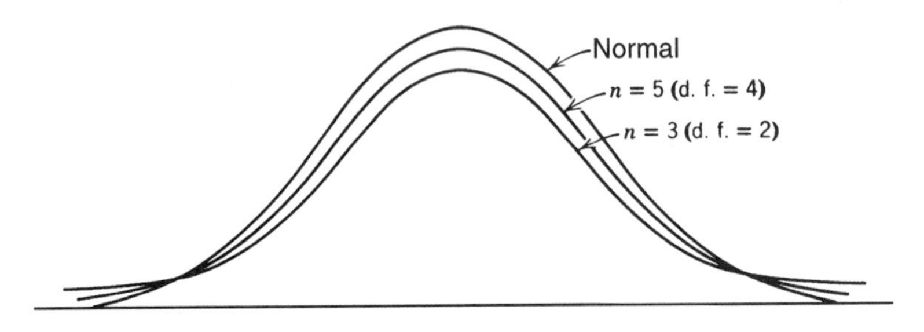

Fig. 6.2. Normal distribution compared with t distributions.

normal distribution, but is somewhat more widely spread than the standard normal distribution. The distribution of t differs for different values of n, the sample size; if the sample size is small, the curve has more area in the "tails"; if the sample size is large, the curve is less spread out and is very close to the standard normal curve.

Figure 6.2 shows the normal distribution and the t distributions for sample sizes $n = 3$ and $n = 5$.

The areas under the distribution curves to the left of t have been computed and put in table form; Table A.3 gives several areas from $-\infty$ to $t[\lambda]$ for some of these distributions. The first column of Table A.3 lists a number called the degrees of freedom (d.f.); this is the number that was used in the denominator in the calculation of s^2, or $n - 1$.

We need to know the degrees of freedom whenever we do not know σ^2. If we knew the population mean μ, the estimate of σ would be $\sum(X - \mu)^2/n$ and the d.f. would be n. When the population mean is unknown we estimate it by the sample mean $\overline{X}$ thus limiting ourselves to samples that have $\overline{X}$ as their sample mean. With such samples, if the first $n - 1$ observations are chosen, the last observation is determined in such a way that the mean of the n observations is $\overline{X}$. Thus we say that there are only $n - 1$ d.f. in estimating s^2.

In the column headings of Table A.3 are the areas from $-\infty$ to $t[\lambda]$, the numbers below which λ% of the area lies. The row headings are d.f. and in the body of Table A.3, $t[\lambda]$ is given. For example, with six d.f. under $\lambda = .90$ we read $t[.90] = 1.440$; 90% of the area under the t distribution lies to the left of 1.440. For each row of the table, the values of $t[\lambda]$ are different depending on the d.f. Note that for d.f.$= \infty$ (infinity), the t distribution is the same as the normal distribution. For example, for $\lambda = .975$ and d.f. $= \infty$, $t[.975] = 1.96$. Also $z[.975] = 1.96$ from Table A.2.

6.4 CONFIDENCE INTERVAL FOR THE MEAN, USING THE t DISTRIBUTION

When the standard deviation is estimated from the sample, a confidence interval for μ, the population mean, is formed just as when σ is known, except that now s replaces σ and the t tables replace the normal tables. With s calculated from the sample, a

95% confidence interval is

$$\overline{X} \pm \frac{t[.975]s}{\sqrt{n}}$$

The $t[.975]$ denotes the value below which 97.5% of the t's lie in the t distribution with $n-1$ d.f.

The quantity $s/\sqrt{n}$ is usually called the *standard error of the mean* in computer program output. Some programs compute confidence limits but with others the user must calculate them using the mean, standard error of the mean from the computer output, and the appropriate t value from either Table A.3 or the computer output.

We return now to the example of the 16 gains in weights with $\overline{X} = 311.9$ and $s = 142.8$ g (see Table 6.1). The standard error of the mean is $142.8/\sqrt{16} = 35.7$. The 95% confidence interval is then

$$311.9 \pm t[.975]\frac{142.8}{\sqrt{16}}$$

The number $t[.975]$ is found in Table A.3 to be 2.131 for d.f. = 15. Thus, 97.5% of the t's formed from samples of size 16 lie below 2.131, and so 95% of them lie between −2.131 and +2.131. This follows from the same reasoning used in discussing confidence intervals when σ was known. The interval is then

$$311.9 \pm 2.131\frac{142.8}{\sqrt{16}}$$

or

$$311.9 \pm 76.1$$

or

$$235.8\text{–}388.0 \quad \text{g}$$

The interpretation of this confidence interval is as follows: We are "95% confident" that μ lies between 235.8 and 388.0 g because, if we keep repeating the experiment with samples of size 16, always using the formula $\overline{X} \pm t[.975]s/\sqrt{n}$ for forming a confidence interval, 95% of the intervals thus formed will succeed in containing μ.

In computing a confidence interval using the t distribution, we are assuming that we have a simple random sample from a normally distributed population. The methods given in Section 5.4 can be used to decide if the observations are normally distributed or if transformations should be considered before computing the confidence interval. Note that in practice we seldom precisely meet all the assumptions; however, the closer we are to meeting them, the more confidence we have in our results.

To find an estimate of the sample size needed for an interval of length L, we must assume a value of σ and proceed as if σ were known. If σ is unknown, an *approximate* estimate can be obtained from the range of likely values for the variable in question using the method given in Section 4.2.2. The difficulty with the use of approximate estimates is that the size of n varies with the square of σ so modest inaccuracies in σ result in larger inaccuracies in n. Sometimes all the researcher can reasonably do is try several likely estimates of σ to gain some insight on the needed sample size.

6.5 ESTIMATING THE DIFFERENCE BETWEEN TWO MEANS: UNPAIRED DATA

In the example used in Section 6.1.1, we estimate the mean gain in weight of infants given a supplemented diet for a 1-month period. Very often, however, our purpose is not merely to estimate the gain in weight under the new diet, but also to find out what difference the new diet supplement makes in weight gained. Perhaps it makes little or no difference, or possibly the difference is really important.

To compare gains in weight under the new supplemented diet with gains in weight under the standard, unsupplemented diet, we must plan the experiment differently. Instead of just giving the entire group of infants the new supplemented diet, we now randomly assign the infants into two groups and give one group (often called the treatment or experimental group) the new supplemented diet and the other group (often called the control group) the standard diet. These two groups are called *independent groups*; here the first group has 16 infants in it and the second group has 9 (see Table 6.1). Whenever possible it is recommended that a control group be included in an experiment.

We compare separate populations; the population of gains in weight of infants who might be given the supplemented diet, and the population of gains in weight of infants who might be given the standard diet. A gain in weight in the *first* population is denoted by X_1, the mean of the first population is called μ_1, and the mean of the first sample is $\overline{X}_1$. Similarly, a member of the *second* population is called X_2, the mean of the second population is called μ_2, and the mean of the second sample is $\overline{X}_2$. The purpose of the experiment is to estimate $\mu_1 - \mu_2$ the difference between the two population means. We calculate $\overline{X}_1$, using the first group of 16 infants and $\overline{X}_2$ from the second group of 9 infants. If $\overline{X}_1 = 311.9$ g and $\overline{X}_2 = 206.4$ g, then the estimate of $\mu_1 - \mu_2$ is $\overline{X}_1 - \overline{X}_2 = 105.5$ g (see Table 6.1). This is the point estimate for $\mu_1 - \mu_2$ and indicates that the new diet may be superior to the old.

6.5.1 The Distribution of $\overline{X}_1 - \overline{X}_2$

After calculating $\overline{X}_1 - \overline{X}_2 = 105.5$ g, we calculate a 95% confidence interval for $\mu_1 - \mu_2$, the difference between the two population means. To calculate the confidence interval, the distribution of $\overline{X}_1 - \overline{X}_2$ is needed. This presents no problem since $\overline{X}_1 - \overline{X}_2$ is simply a statistic that can be computed from the experiment. If the experiment were repeated over and over, the value of $\overline{X}_1 - \overline{X}_2$ would vary from one experiment to another and thus it has a sampling distribution.

It is usual to assume in this problem that the variances of the two original populations are equal; the two equal variances will be called σ^2. If the first population is approximately normally distributed, we know that $\overline{X}_1$ is approximately normally distributed, and that its mean is μ_1, and its standard deviation is $\sigma/\sqrt{n_1} = \sigma/\sqrt{16} = \sigma/4$ g, where $n_1 = 16$ is the size of the first group of infants. Similarly, $\overline{X}_2$ is approximately normally distributed with mean μ_2 and standard deviation

$$\sigma/\sqrt{n_2} = \sigma/\sqrt{9} = \sigma/3 \quad \text{g}$$

where $n_2 = 9$ is the sample size of the second group of infants.

If all possible pairs of samples were drawn from the two populations and for each pair $\overline{X}_1 - \overline{X}_2$ is calculated, what kind of distribution would this statistic have? First, the distribution is approximately normal; this does not seem surprising. Second, the mean of the distribution is $\mu_1 - \mu_2$; this seems reasonable. Third, the standard deviation is larger than the standard deviation of the $\overline{X}_1$ distribution and is also larger than that of the $\overline{X}_2$ distribution; in fact, the variance for $\overline{X}_1 - \overline{X}_2$ is equal to the sum of the two variances. This may be somewhat surprising; evidence to support it can be obtained in a class exercise with the samples of cholesterol levels from Problem 2.1. At any rate, it has been proven mathematically that

$$\sigma^2_{\overline{X}_1-\overline{X}_2} = \sigma^2_{\overline{X}_1} + \sigma^2_{\overline{X}_2} = \frac{\sigma^2}{n_1} + \frac{\sigma^2}{n_2} = \sigma^2\left(\frac{1}{n_1} + \frac{1}{n_2}\right)$$

Note that when the sample sizes of the two groups are equal, $n_1 = n_2$ or simply n, that $\sigma^2_{\overline{X}_1-\overline{X}_2} = 2\sigma^2/n$.

Figure 6.3 illustrates the appearance of the distributions of $\overline{X}_1$, $\overline{X}_2$, and $\overline{X}_1 - \overline{X}_2$.

6.5.2 Confidence Intervals for $\mu_1 - \mu_2$: Known Variance

It can be shown mathematically, then, that for independent samples of size n_1 and n_2 from normal populations with means μ_1 and μ_2 and with equal variances σ^2, the statistic $\overline{X}_1 - \overline{X}_2$ has a normal distribution with mean $\mu_1 - \mu_2$ and with variance

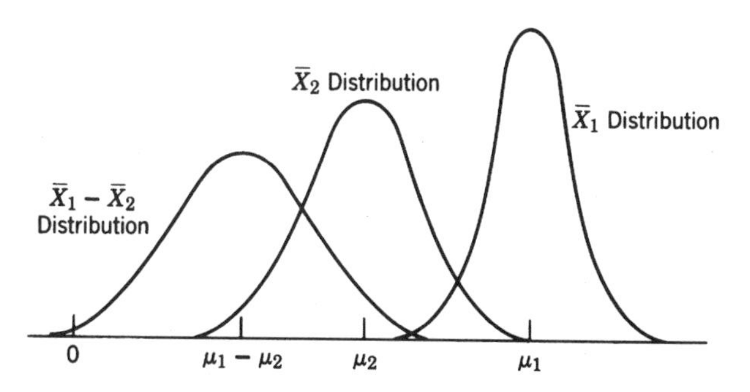

Fig. 6.3. Two distributions of sample means compared with distribution of their differences.

$\sigma^2(1/n_1 + 1/n_2)$. In Section 6.1.2, when we had $\overline{X}$ normally distributed with mean μ and with variance σ^2/n, and σ^2 was known, the 95% confidence interval for μ was $\overline{X} \pm 1.96\sigma/\sqrt{n}$. Analogous to this, if σ^2 is known, a 95% confidence interval for $\mu_1 - \mu_2$ is

$$(\overline{X}_1 - \overline{X}_2) \pm 1.96\sigma\sqrt{\frac{1}{n_1} + \frac{1}{n_2}}$$

For $\overline{X}_1 = 311.9$ g, and $\overline{X}_2 = 206.4$ g, if it is known that $\sigma = 120$ g, the interval is

$$(311.9 - 206.4) \pm 1.96(120)\sqrt{\tfrac{1}{9} + \tfrac{1}{16}}$$

or

$$105.5 \pm 1.96(120)\sqrt{0.1736}$$

or

$$105.5 \pm 1.96(120)(0.417)$$

or

$$105.5 \pm 98.1$$

or

$$7.4\text{–}203.6 \quad \text{g}$$

Both confidence limits are positive, and so we conclude that the supplemented diet does increase weight gains on the average in the population (i.e., we conclude that $\mu_1 - \mu_2$ is positive and $\mu_1 > \mu_2$). It is also obvious from the length of the confidence interval ($203.6 - 7.4 = 196.2$ g) that the size of the difference covers a wide range.

Here, we are assuming that we have simple random samples from two independent populations that both have a variance σ^2 and that the computed z is normally distributed.

To estimate the sample size needed for a confidence interval of length L when $n_1 = n_2$ or simply n, we replace σ with $\sqrt{2}\sigma$ in the formula for sample size for a single group.

6.5.3 Confidence Intervals for $\mu_1 - \mu_2$: Unknown Variance

Usually, the population variance σ^2 is unknown and must be approximated by some estimate that can be calculated from the samples. An estimate s_1^2 can be calculated from the first sample using $s_1^2 = \sum (X_1 - \overline{X}_1)^2/(n_1 - 1)$ and similarly an estimates

s_2^2 can be computed from the second sample. These two estimates are then pooled to obtain the pooled estimate of the variance, s_p^2 . The pooled estimate of the variance is a weighted mean of s_1^2 and s_2^2 and its formula is

$$s_p^2 = \frac{(n_1 - 1)s_1^2 + (n_2 - 1)s_2^2}{n_1 + n_2 - 2}$$

When $\overline{X}_1 - \overline{X}_2$ are normally distributed with mean $\mu_1 - \mu_2$, and variance $\sigma^2(1/n_1 + 1/n_2)$, then the quantity

$$z = \frac{(\overline{X}_1 - \overline{X}_2) - (\mu_1 - \mu_2)}{\sigma\sqrt{1/n_1 + 1/n_2}}$$

has a standard normal distribution. Substituting s_p for σ gives a quantity whose distribution is a t distribution. The d.f. for this t distribution are $n_1 + n_2 - 2$, which is the number used in the denominator of s_p^2. [Recall that for a single sample, $t = (\overline{X} - \mu)/(s/\sqrt{n})$ had $n - 1$ d.f., and that $n - 1$ is the number used in the denominator of s^2.]

The 95% confidence interval for $\mu_1 - \mu_2$ may then be computed by substituting s_p for σ and by changing 1.96, the 97.5% point of the z distribution to the 97.5% point of the t distribution with $n_1 + n_2 - 2$ d.f. That is, the interval becomes

$$(\overline{X}_1 - \overline{X}_2) \pm t[.975]s_p\sqrt{1/n_1 + 1/n_2}$$

with d.f. $= n_1 + n_2 - 2$. The calculation of the 95% confidence interval with σ^2 unknown can be illustrated for $\mu_1 - \mu_2$. Here, $n_1 = 16$, $n_2 = 9$, $\overline{X}_1 = 311.9$ g, $\overline{X}_2 = 206.4$ g, $s_1^2 = 20{,}392$, $s_2^2 = 7060$, and $t[.975] = 2.069$ for $n_1 + n_2 - 2 = 23$ d.f. First, we compute the pooled variance as

$$s_p^2 = \frac{(16-1)(20,392) + (9-1)(7060)}{16+9-2} = \frac{362,360}{23} = 15,755$$

Taking the square root of the pooled variance yields $s_p = 125.5$. The confidence interval for $\mu_1 - \mu_2$ for σ unknown is

$$(311.9 - 206.4) \pm (2.069)(125.5)\sqrt{\tfrac{1}{16} + \tfrac{1}{9}}$$

or

$$105.5 \pm 2.069(125.5)\sqrt{0.1736}$$

or

$$105.5 \pm 108.2$$

or the interval is from −2.7 to 213.7 g. Because the lower limit is negative and the upper limit is positive, we cannot conclude that the population mean gain in weight under the supplemented diet is larger than under the standard diet; the difference $\mu_1 - \mu_2$ may be either positive or negative but we might be inclined to think that it is positive.

In the past example, we have assumed that we have simple random samples from two normal populations with equal population variances. Methods for determining if the data are approximately normally distributed were discussed in Section 5.4. A method for testing if the two variances are equal will be given in Section 10.3.

6.6 ESTIMATING THE DIFFERENCE BETWEEN TWO MEANS: PAIRED COMPARISON

Sometimes in studying the difference between two means it is possible to use pairs or matched samples advantageously. This device is often quite effective in partially eliminating the effects of extraneous factors. As mentioned in Chapter 2, matching is commonly used in case/control studies where patients are matched by gender and age. It is often used in surgical experiments where two surgical treatments are used on the opposite sides of animals such as rats to compare the outcomes. Another common use of paired comparisons is to contrast data gathered in two time periods.

For example, consider data on eight adults before and after going on a diet given in Table 6.2. We will let X_1 denote the weight before dieting (Time 1) and X_2 the weight after dieting for 2 months (Time 2). The difference in the before and after weight can be computed for each adult as $d = X_1 - X_2$ and there are eight differences. We will assume that we have a simple random sample of differences and that the differences are normally distributed.

We now analyze these differences instead of the 16 original weights. We reduced the problem from one of two sets of observations on weights to a single set of dif-

Table 6.2. Weight Loss in lbs. between Time 1 and Time 2 for Eight Adults

Adults Number	Time 1 X_1	Time 2 X_2	Difference d
1	278	271	7
2	183	181	2
3	175	175	0
4	199	190	9
5	210	204	6
6	167	164	3
7	245	239	6
8	225	221	4
$n = 8$	$\overline{X}_1 = 210.25$	$\overline{X}_2 = 205.625$	$\overline{d} = 4.625$
	$s_1 = 37.8$	$s_2 = 36.2$	$s_d = 2.92$

ferences. We now treat the d's as eight observations and first compute the mean and standard deviation using a statistical package. We obtain

$$\bar{d} = \frac{\sum d}{n} = 4.625 \quad \text{and} \quad s_d = \sqrt{\frac{\sum(d - \bar{d})^2}{n - 1}} = 2.9246$$

Then, the 95% confidence interval can be computed using the formula given earlier for a single mean where the X's are replaced by d's. There are 8 differences and $n - 1 = 7$ d.f. The 95% interval is

$$\bar{d} \pm t[.975]\frac{s_d}{\sqrt{n}}$$

or

$$4.625 \pm (2.365)\frac{2.9246}{\sqrt{8}}$$

or

$$4.625 \pm 2.445 \quad \text{or} \quad 2.18\text{–}7.07 \quad \text{lb}$$

The interval contains only positive numbers and we conclude that μ_d is positive and that, on the average, the diet decreases weight for the adults.

In this example, the pairing of the weights on the same adult was a natural outcome of how the measurements were taken; it is obvious that the paired comparison is the one to use. When artificial pairs are created in clinical trials to assure that the treatment and control group are similar in say age and gender, the reduction in the length of the confidence interval is often not great. Since patients usually enter the trial one at a time, often the investigator may be unable to find a close match for a patient already entered. Random assignment is strongly recommended for clinical trials. In case/control studies, since random assignment is not possible, pairing is often done not so much to reduce the size of the confidence interval as to insure that the cases and controls are similar on some factors that might affect the outcome. The decision to pair or not to pair often depends on the type of research study being done and whether it is a simple process to find pairs.

In general, without a practical, relatively easy way of pairing, it may be better to use independent groups and make the number of observations somewhat larger.

PROBLEMS

6.1 Make a 95% confidence interval for the mean of the cholesterol level population using each one of the samples gathered in Exercise 2.1, assuming that you know that $\sigma = 45.3$ mg/100 mL (three confidence intervals, one for each sample).

6.2 Gather all the confidence intervals that the class constructed in Exercise 6.1, and make a graph plotting each interval as a vertical line. The population mean of the 98 cholesterol measurements is 279.5 mg/100 mL. How many intervals cross a horizontal line $\mu = 279.5$ mg/100 mL? About how many would you expect to cross a line $\mu = 279.5$ mg/100 mL?

6.3 The time taken for cessation of bleeding was recorded for a large number of persons whose fingers had been pricked. The mean time was found to be 1.407 min and the standard deviation was 0.588 min. In an effort to determine whether pressure applied to the upper arm increases bleeding time, six persons had pressure equal to 20 mm applied to their upper arms and their fingers pricked. For these six persons, the times taken for bleeding to stop were 1.15, 1.75, 1.32, 1.28, 1.39, and 2.50 min.

Give a 95% confidence interval for the mean bleeding time under pressure for the six persons and draw some conclusion as to whether pressure increases bleeding time.

6.4 Mice from 12 litters were weighed when 4-months old, with one male and one female weighed from each litter. Weight in grams was recorded as follows:

Litter Number	Male	Female
1	26.0	16.5
2	20.0	17.0
3	18.0	16.0
4	28.5	21.0
5	23.5	23.0
6	20.0	19.5
7	22.5	18.0
8	24.0	18.5
9	24.0	20.0
10	25.0	28.0
11	23.5	19.5
12	24.0	20.5

Give a 95% confidence interval for the difference between the mean weight of male mice and female mice.

6.5 To study whether or not pressure exerted on the upper arm increases bleeding time, 39 persons had their upper arm subjected to a pressure of 40 mmHg of mercury and their fingers pricked. Their mean bleeding time was 2.192 min, and the standard deviation of their bleeding times was 0.765 min. Forty-three other persons acted as controls. No pressure was used for the controls, and their bleeding time was found to have a mean of 1.524 minutes and a standard deviation of 0.614 min.

Give a 95% confidence interval for the difference in bleeding times. Do you think pressure increases mean bleeding time? Why?

6.6 **(a)** A special diet was given to 16 children and their gain in weight over a 3-month period was recorded. Their mean gain in weight was found to be 2.49 kg. A control group, consisting of 16 children of similar background and physique, had normal meals during the same period and gained 2.05 kg on the average.

Assume that the standard deviation for weight gains is 0.8 kg. Is the evidence strong enough for us to assert that the special diet really promotes gain in weight?

(b) Answer the question in (a) if each of the two groups consisted of 50 children.

6.7 A general physician recorded the oral and rectal temperatures of nine consecutive patients who made first visits to his office. The temperatures are given in degrees Celcius (°C). The following measurements were recorded:

Patient Number	Oral Temperature (°C)	Rectal Temperature (°C)
1	37.0	37.3
2	37.4	37.8
3	38.0	39.3
4	37.3	38.2
5	38.1	38.4
6	37.1	37.3
7	37.6	37.8
8	37.9	37.9
9	38.0	38.3

(a) From this data, what is your best point estimate of mean difference between oral and rectal temperatures?

(b) Give a 95% confidence interval for this difference and state what the interval means.

(c) Give a 95% confidence interval for mean oral temperature.

(d) Give a 95% confidence interval for mean rectal temperature.

6.8 In an experiment reported in the *Lancet* on the effectiveness of isoniazid as a treatment for leprosy in rats, the log survival times (weeks) were evaluated. For untreated rats, the log time was 1.40, 1.46, 1.48, 1.52, 1.53, 1.53, 1.56, 1.58, 1.62, and 1.65. For treated rats, it was 1.71, 1.72, 1.83, 1.85, 1.85, 1.86, 1.86, 1.90, 1.92, and 1.95. The rats were randomly assigned to the two groups so the groups are independent. If log times in the untreated group are called X_1 and in the treated group are called X_2, then the sample means and sums of squares of deviations are $\overline{X}_1 = 1.533$, $\overline{X}_2 = 1.845$, $\sum(X_1 - \overline{X}_1)^2 = 0.050$, and $\sum(X_2 - \overline{X}_2)^2 = 0.054$.

Calculate (a) s_p; (b) the degrees of freedom; (c) a 99% confidence interval for the true mean difference in log units; and (d) on the basis of the interval obtained in (c), do you feel reasonably sure that the treatment was effective? Why?

6.9 Bacterial colonies were counted on 12 plates. There were two observers, and the plate counts obtained were as follows:

Plate Number	Observer 1	Observer 2	Difference
1	139	191	−52
2	121	181	−60
3	49	67	−18
4	163	143	20
5	191	234	−43
6	61	80	−19
7	179	250	−71
8	218	239	−21
9	297	289	8
10	165	201	−36
11	91	80	11
12	92	99	−7

Give a 99% confidence interval for the difference between the mean plate counts of the two observers. Do you believe there is a difference? Why?

6.10 A physician is planning to take systolic blood pressure measurements from a group of office workers and a group of repair personnel working for a telephone company. Past results indicated that σ is approximately 15 mmHg. If it is desired to have a 95% confidence interval of length 4 mmHg, how many workers should be measured in each group?

REFERENCES

Dunn, O. J. and Clark, V. A. [1987]. *Applied Statistics: Analysis of Variance and Regression,* New York: Wiley, 40–43.

Fisher, L. D. and van Belle, G. [1993]. *Biostatistics: A Methodology for the Health Sciences,* New York: Wiley-Interscience, 103–105.

Rothman, K. J. [1986]. *Modern Epidemiology,* Boston, MA: Little, Brown, and Company, 119–125.

CHAPTER SEVEN

Tests of Hypotheses on Population Means

In Chapter 6, we presented the methods used in constructing confidence intervals for the population mean, a widely reported parameter. This chapter also considers population means but here we show how to test hypotheses concerning them. Hypothesis testing is a commonly used method for assessing whether or not sample data are consistent with statements made about the population. In addition to this chapter, hypothesis testing will also be discussed in Chapters 8–11.

In Section 7.1, hypothesis testing is introduced by presenting an example for a single mean. Tests for two means using data from independent populations are given in Section 7.2, and tests for paired data are given in Section 7.3. The general concepts used in testing hypotheses are explained in Section 7.4. The needed sample size for tests of one or two means are given in Section 7.5. Sections 7.6–7.8 discuss comparisons between the use of confidence intervals and hypothesis testing, correction for multiple testing, and reporting the results.

7.1 TESTS OF HYPOTHESES FOR A SINGLE MEAN

In this section, we present tests that are used when only one mean is being tested; one test for when the population standard deviation σ is known and one where it is unknown. We also discuss the usual procedures that are followed in tests concerning means.

7.1.1 Test for a Single Mean when σ Is Known

In a study of children with congenital heart disease, researchers gathered data on age in months of walking from 18 children with acyanotic congenital heart disease; the data are recorded in Table 7.1. From larger studies on normal children, the researchers know that mean age of walking for normal children is 12.0 months and that the standard deviation is 1.75 months. We wish to decide whether or not acyanotic children learn to walk at the same age on the average as normal children. We formulate a hypothesis

Table 7.1. Age of Walking for 18 Acyanotic Children

Child	Age (months)	Child	Age (months)
1	15.0	10	10.2
2	11.0	11	14.0
3	14.2	12	14.8
4	10.0	13	14.2
5	12.0	14	16.0
6	14.2	15	13.5
7	14.5	16	9.2
8	13.8	17	15.0
9	12.8	18	16.5
	$\overline{X} = 13.38$		$s = 2.10$

that we propose to test. This proposed hypothesis is called a null hypothesis. Our null hypothesis is as follows: The mean age of walking for a population of acyanotic children is 12.0 months. The null hypothesis is equivalent to saying that acyanotic children learn to walk at the same age on the average as normal children. On the basis of the data, we wish either to *accept the null hypothesis* or to *reject the null hypothesis*. The null hypothesis should be decided upon prior to the analysis of the data.

The null hypothesis to be tested is often written in symbols, here H_0: $\mu = \mu_0$. H_0 denotes the null hypothesis, μ denotes the mean of the population being studied, and μ_0 denotes the hypothesized mean. In the acyanotic children example, the null hypothesis is H_0: $\mu = 12.0$. At this point, the student might suggest that we compute $\overline{X}$ from the sample and see whether or not it is equal to 12.0. But because of sampling variation, even if the average age of walking for the population of acyanotic children is 12.0 months, this particular sample of 18 children may have $\overline{X}$ either higher or lower than 12.0 months. What must be decided is whether $\overline{X}$ is *significantly different* from 12.0.

We reason this way: There is a whole population of acyanotic children; its mean age of walking is unknown, but its standard deviation is known to equal 1.75 months. (Here we assume that the variability in the age of walking is the same in acyanotic children as in well children.) We pretend temporarily that the population mean age of walking for acyanotic children is 12 months, and decide whether the sample of size 18 could easily have been drawn from such a population. A simple random sample of children with acyanotic congenital heart disease is assumed.

Under the temporary assumption, then, that $\mu = 12.0$, the means of repeated samples of size 18 would form a population whose mean is also 12.0 months and whose standard deviation is $\sigma_{\overline{X}} = 1.75/\sqrt{18} = 0.412$ months. The population of $\overline{X}$'s is approximately normally distributed provided the original population is not far from normal.

We compute $\overline{X} = 13.38$ months (Table 7.1). The question is, "How unusual is this sample mean, assuming that the sampling was done from a population whose mean was 12.0 months?" To answer this, we compute the chance that a sample be drawn whose sample mean is at least as far from $\mu = 12.0$ as is $\overline{X} = 13.38$. In other words, we compute the probability of sample means occurring that are as unusual or more unusual than this one. Since the null hypothesis is H_0: $\mu = 12.0$, we reject the null hypothesis if the computed $\overline{X}$ is either *much smaller* than 12.0 or *much larger*.

This probability will be called P; it can be obtained by finding the area lying above $\overline{X} = 13.38$ under the normal curve whose mean is 12.0 and whose population standard deviation of the mean is 0.412, and also the area lying below 10.62. (The point 10.62 is as far below 12.0 as the point 13.38 is above 12.0. Since the normal curve is symmetric, all sample means that are at least as unusual as 13.38 lie either below 10.62 or above 13.38.) To find the z value for a $\overline{X} = 13.38$, $\mu = 12.0$, and $\sigma_{\overline{X}} = 0.412$, we compute

$$z = \frac{\overline{X} - \mu_0}{\sigma/\sqrt{n}}$$

or

$$z = (13.38 - 12.0)/0.412 = 1.38/0.412 = 3.35$$

By looking in Table A.2, we find for $z = 3.35$ that the probability of obtaining a sample mean corresponding to $z < 3.35$ is .9996, so that the chance of a sample mean > 13.38 is only .0004 or 0.04%. The probability of obtaining a sample mean such that z is less than -3.35 is also equal to .0004 (see Fig. 7.1). We will add these two probabilities to obtain .0008. So, if 12.0 is the population mean, a sample mean as unusual as 13.38 would be obtained on the average, in repeated sampling, only 8 times in 10,000 (in other words, P is merely .0008 or 0.08%). Thus, either a very unusual sample has been drawn, or else the population mean is not 12.0. It is of course possible that the population mean is really 12.0 and that an unusual sample was drawn, but it seems more likely that the population mean is simply not 12.0. We decide *to reject the null hypothesis*. That is, we conclude that the mean age at walking for acyanotic children differs from 12.0 months, the mean for well children.

It is possible that our conclusion is incorrect, and that the mean age is actually 12.0, and that this sample is an unusual one. We cannot *know* that μ is not 12.0, since we have only one sample from the population on which to base the decision. However,

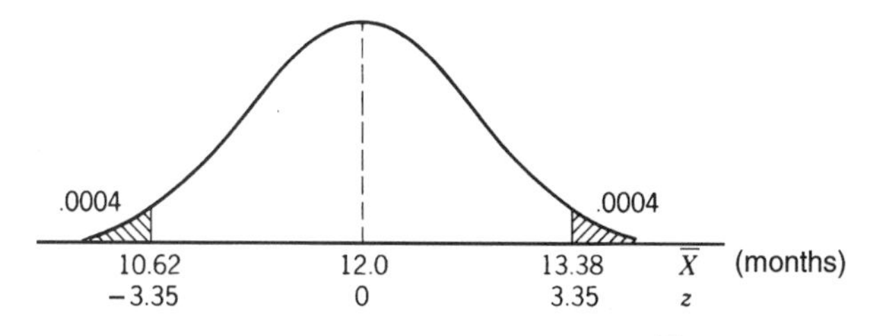

Fig. 7.1. Two-sided test of $H_0 : \mu = 12.0$, when $\overline{X} = 13.38$.

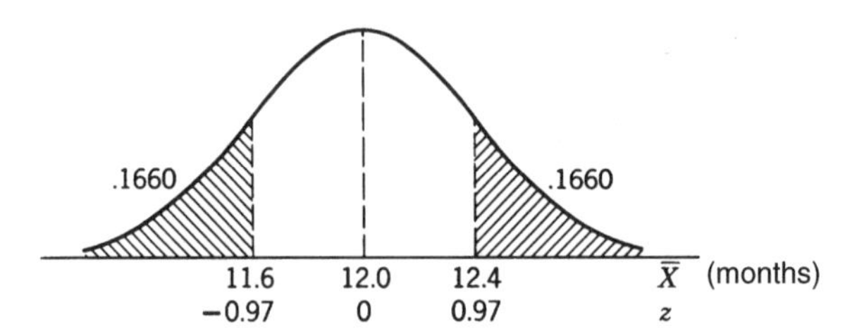

Fig. 7.2. Two-sided test of $H_0 : \mu = 12.0$, when $\overline{X} = 11.6$.

the conclusion that μ is not 12.0 seems more reasonable than that a highly unusual sample was obtained.

Suppose that we had obtained $\overline{X} = 11.6$ months for the sample mean instead of $\overline{X} = 13.38$. We proceed exactly as we did with $\overline{X} = 13.38$. This time z is

$$z = (11.6 - 12.0)/0.412 = -0.4/0.412 = -0.97$$

In Table A.2, the area to the left of $z = +0.97$ is .8340, or 83.4%. Note that even though the computed $z = -0.97$ we use $z = 0.97$. This is possible because the normal distribution is symmetric. The chance of a sample mean's being < 11.6 is then $1.00 - .8340 = .1660$, and the chance of a sample mean's being < 11.6 or > 12.4 (or 0.4 units from $\mu = 12.0$) is twice as large, so that $P = 2(.1660) = .3320$ or 33% (see Fig. 7.2).

In other words, if the population mean is really 12.0, and if the experiment were repeated many times, and each time the mean age of walking calculated from a random sample of 18 acyanotic children, about 33 out of 100 of the sample means would differ from 12.0 by at least as much as did the sample mean actually obtained. Clearly, 11.6 is close enough to 12.0 that it could have been obtained very easily by chance. We look at $P = .33$ and decide to accept the null hypothesis. The acceptance is a negative type of acceptance; we simply have failed to disprove the null hypothesis. We decide that, as far as we can tell from the sample, the population mean may be 12.0 and that acyanotic heart disease children may learn to walk at the same average age as other children. It should be emphasized that this conclusion may be incorrect; the mean may actually be something different from 12.0. But in any case, a sample mean of 11.6 could easily have arisen from a population whose mean was 12.0, so if we still wish to establish that the age of walking is different for acyanotic children from that for normal children, they must gather more data.

In these examples, it was fairly easily decided to reject the null hypothesis when P was .0008, and similarly it was easily decided to accept the null hypothesis when P was equal to .33. It is not so apparent which decision to make if we obtain a P somewhere in between, say $P = .1$, for example. Where should the dividing line be between the values of P that would lead us to reject the null hypothesis and those that would lead us to accept the null hypothesis? In much experimental work, the dividing line is chosen in advance at .05. If P is $< .05$ and if the population mean is 12.0, then the chance is < 1 in 20 that we will have a sample mean as unusual as the 13.38 months occur; we therefore conclude that the mean is not 12.0.

In statistical literature, the number picked for the dividing line is called α (alpha); it is seen as our chance of making the mistake of deciding that the null hypothesis is false when it is actually true. Alpha is sometimes called the *level of significance*. It is advisable to choose the level of α before obtaining the results of the test so that the choice is not dependent on the numerical results. For example, when α is chosen to be .05 in advance, then we only have to see if the *absolute* value of the computed z value was $> z[.975] = 1.96$ for H_0: $\mu = \mu_0$. When the null hypothesis is rejected, the results are said to be *statistically significant*. If the computed z is > 1.96 or < -1.96, we say that the value of z lies in the *rejection region* and the null hypothesis is rejected. If the value of z lies between -1.96 and $+1.96$, then it lies in the *acceptance region*, and the null hypothesis is not rejected. If the null hypothesis is not rejected, then the results are often stated to be *nonsignificant*.

7.1.2 One-Sided Tests when σ Is Known

In the previous example concerning age of walking for acyanotic children, we wished to find out whether or not the mean is 12.0; the test that they made is called a two-sided test because computing P included the chance of getting a sample mean above the one actually obtained (13.38) and also the chance of getting a sample mean < 10.62.

Sometimes it is appropriate, instead, to make a one-sided test. We may consider it highly unlikely that acyanotic children learn to walk earlier on the average than normal children and may not be interested in rejecting the null hypothesis if this occurs. Then, the question to be answered by the experiment is no longer whether the mean is 12.0, or something else. Instead, the question is whether the mean is > 12.0 or not. We do not want to reject the null hypothesis for small values of the mean. Then, to calculate P we find the proportion of sample means that lie above $\overline{X}$. If from the sample, $\overline{X}$ is calculated to be 13.38, then P from Table A.2 is found to be .0004.

The null hypothesis is stated H_0: $\mu \leq 12$ or in general H_0: $\mu \leq \mu_0$. If the null hypothesis is rejected, then we conclude that the population mean is > 12.0, that is, acyanotic congenital heart disease children learn to walk at an older age than normal children. The entire rejection region is in the upper tail of the normal distribution. For $\alpha = .05$, $z[.95] = 1.65$, and the null hypothesis is rejected if z is > 1.65.

Sometimes we only wish to reject the hypothesis if the population mean is too small. For example, suppose that we are testing the weight of a medication; the producers say that the bottles hold medication weighing 6 oz. We would only want to reject their claim if the bottles weigh too little. Here, the entire rejection region is in the lower tail. We would state our null hypothesis as H_0: $\mu \geq 6$ and reject the null hypothesis if $z \leq -1.65$ if we wish to test at an $\alpha = .05$ level. Note that for $z[-1.65]$ we have by symmetry 5% of the area in the lower or left tail of the normal distribution.

7.1.3 Summary of Procedures for Test of Hypotheses

The usual steps that are taken in testing a null hypothesis include

1. State the purpose of making the test. What question is to be answered?
2. State the null hypothesis to be tested (two sided or one sided).
3. Choose a level of α that reflects the seriousness of deciding that the null hypothesis is false when actually it is true.
4. Decide on the appropriate *test statistic* to use for testing the hypothesis. Here, since we are assuming σ is known, the test statistic is $z = (\overline{X} - \mu_0)/\sigma_{\overline{X}}$.
5. Check that the sample and data meet the assumptions for the test. In this case, we are assuming a simple random sample taken from a population that has σ equal to the assumed value. We are assuming that z follows a normal distribution. We should check that there are no errors or outliers in our sample and that the distribution is sufficiently close to normal so that the sample mean is approximately normally distributed. Also, we may wish to check whether the numerical value of s is reasonably close to the assumed σ (see Section 10.2.1 for a discussion of this topic).
6. Compute the value of the test statistic, in this case

$$z = \frac{(\overline{X} - \mu_0)}{\sigma/\sqrt{n}}$$

7. Either find the tabled values for α and check the computed test statistic against the tabled values or obtain the P value for the numerical value of the test statistic. For example, in this case for a two-sided test and $\alpha = .05$, see if the computed z is > 1.96 or < -1.96. Otherwise, take the computed z value and look up the area in Table A.2 or in a computer program. Double the area found in one tail to obtain the P value for a two-sided test.
8. State the statistical conclusion either to reject or fail to reject the null hypothesis. This should be followed with a statement of the results as they apply to the goals of the research project.

7.1.4 Test for a Single Mean when σ Is Unknown

In the earlier example of 18 children with acyanotic heart disease, the standard deviation of the original population was assumed to be 1.75 months, the same standard deviation as for age of walking for well children. The assumption that the variability of a population under study is the same as the known variability of a similar population is sometimes reasonable. More often, however, the standard deviation must be estimated from the sample data, either because one does not know the standard deviation of the comparable population (one does not know that 1.75 months is the standard deviation for age of walking for normal children) or because one is not sure that the variability of the population being studied is close to that of the comparable population (a small proportion of children with heart disease may walk very late, so that age of walking for acyanotic children may be more variable than for normal children). Just as in obtaining confidence intervals in Section 6.4, s must be calculated from the sample and used in place of σ, and the t distribution must be used instead of the normal distribution. Here, the $\overline{X} = 13.38$ and $s = 2.10$ (see Table 7.1). The standard error of the mean is $s_{\overline{X}} = s/\sqrt{n}$ or $2.10/\sqrt{18} = 0.4944$. The test statistic is then

$$t = \frac{(\overline{X} - \mu_0)}{s_{\overline{X}}}$$

or

$$t = \frac{(13.38 - 12.0)}{0.4944} = 2.79$$

If the test is being made to decide whether μ is > 12.0, a one-sided test is called for. The null hypothesis is H_0: $\mu \leq 12.0$, so if it is rejected the conclusion is that the population mean is > 12.0. In Table A.3, we enter the table with 17 d.f. and look for 2.79; we find 2.567 in the column headed .99 and 2.898 in the column headed .995. Thus the area to the left of $t = 2.79$ is between .99 and .995 and the area to the right of 2.79 is between .01 and .005. Thus $.005 < P < .01$. Usually, this is stated simply as $P < .01$ (see Fig. 7.3).

If we have chosen $\alpha = .05$ as the level of significance, we reject the null hypothesis and decide that the mean age of walking is > 12.0 months; acyanotic children walk later on the average than normal children. If we wish to test the null hypothesis at the $\alpha = .05$ level without computing P, we compare the calculated t=2.79 with the tabled value $t[.95] = 1.74$ with 17 d.f. and reject the null hypothesis because 2.79 is > 1.74.

If a two-sided test is performed, then the hypothesis is H_0: $\mu = 12.0$ and is rejected if the sample mean is either too small or too large. The area to the right of 2.79 is doubled to include the lower tail and thus $.01 < P < .02$, again the test would be rejected. Alternatively, the computed value of $t = 2.79$ can be compared with the tabled value $t[.975] = 2.11$ with 17 d.f. and since $2.79 > 2.11$, the null hypothesis of equal means would be rejected. Using a two-sided test with $\alpha = .05$ and $t[.975]$ implies that one-half of the rejection region is in the upper or right tail and one-half is in the lower tail. In making this t test, we are assuming that we have a simple random

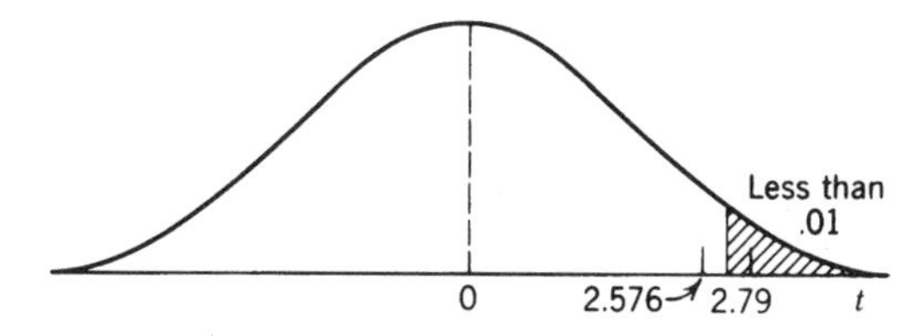

Fig. 7.3. One-sided test of $H_0 : \mu \leq 12.0$.

sample and that the observations are normally distributed. The normality assumption can be checked using the methods given in Section 5.4. If the data are not normally distributed, transformations can be considered (see Section 5.5).

Computer programs usually used do not include the test using a known σ, but automatically use the calculated standard deviation and perform the t test. The t test is often called the Student t test. The programs usually provide the results only for two-sided tests.

Some programs will allow the user to test for any numerical value of μ_0 such as $\mu_0 = 12.0$. Some programs will allow the user to only test that $\mu_0 = 0$. When using one of these programs, the user should have the program create a new variable by subtracting the hypothesized μ_0 from each of the original observations. In the example, the new variable would have observations that are $X - 12.0$. The t test is then performed using the new variable and H_0: $\mu = 0$. One of the advantages of using computer programs is that they can perform the calculations and give accurate P values for any d.f. In other words, the user does not have to say that the P value is between two limits such as $.01 < P < .02$ or that $P < .02$ but instead can say that P equals a particular value.

7.2 TESTS FOR EQUALITY OF TWO MEANS: UNPAIRED DATA

As a second example of testing a hypothesis, let us suppose that we are interested in comparing the hemoglobin level of children with *acyanotic* heart disease with that of children with *cyanotic* heart disease. The question is, "Is the mean hemoglobin level for acyanotic children different from that for cyanotic children?" The null hypothesis is that the mean levels for acyanotic and cyanotic are the same.

7.2.1 Testing for Equality of Means when σ Is Known

From experience with hemoglobin levels, researchers expect the population standard deviation to be about 1.0 g/cc. The available data consists of two samples, one of 19 acyanotic children, the other of 12 cyanotic children (see Table 7.2). The mean for each sample is computed as $\overline{X}_1 = 13.03$ g/cc for the acyanotic children and as $\overline{X}_2 = 15.74$ for the cyanotic children. We calculate the difference between them, $\overline{X}_1 - \overline{X}_2 = -2.71$ g/cc. The problem then is to decide whether there really is a difference between the mean hemoglobin levels of the two populations of children with congenital heart disease or whether the difference $\overline{X}_1 - \overline{X}_2 = -2.71$ was caused simply by sampling variation. (One way of stating this question is to ask whether the difference between $\overline{X}_1$ and $\overline{X}_2$ is *significant*.)

We have two populations: A population of hemoglobin levels of acyanotic children (X_1's), with mean μ_1, and a population of hemoglobin levels of cyanotic children (X_2's), with mean μ_2, and it is assumed that the two populations have the same population standard deviations of 1.0 g/cc.

From Section 6.5, the population of $\overline{X}_1 - \overline{X}_2$'s obtained taking all possible samples of size n_1 and n_2, respectively, from the two populations, has a mean $\mu_1 - \mu_2$ and a

Table 7.2. Hemoglobin Levels for Acyanotic and Cyanotic Children

Acyanotic		Cyanotic	
Number	X_1 (g/cc)	Number	X_2 (g/cc)
1	13.1	1	15.6
2	14.0	2	16.8
3	13.0	3	17.6
4	14.2	4	14.8
5	11.0	5	15.9
6	12.2	6	14.6
7	13.1	7	13.0
8	11.6	8	16.5
9	14.2	9	14.8
10	12.5	10	15.1
11	13.4	11	16.1
12	13.5	12	18.1
13	11.6		
14	12.1		
15	13.5		
16	13.0		
17	14.1		
18	14.7		
19	12.8		
	$\overline{X}_1 = 13.03$ $s_1^2 = 1.0167$		$\overline{X}_2 = 15.74$ $s_2^2 = 1.9898$

standard deviation

$$\sigma\sqrt{1/n_1 + 1/n_2} = 1.0\sqrt{1/19 + 1/12} = 1.0\sqrt{0.13596} = 0.3687 \text{ g/cc}.$$

The hypothesis that will be tested is H_0: $\mu_1 = \mu_2$; it may also be written as H_0: $\mu_1 - \mu_2 = 0$, which is equivalent to saying that $\mu_1 - \mu_2$, the mean of the $\overline{X}_1 - \overline{X}_2$ population, is 0. If the original populations are not far from normal and simple random sampling can be assumed, the $\overline{X}_1 - \overline{X}_2$ population is approximately normally distributed. Under the null hypothesis, then, we have a population of $\overline{X}_1 - \overline{X}_2$'s, which is normally distributed, whose mean is 0, and whose standard deviation is 0.3687. This hypothesis will be rejected if the numerical value of $\overline{X}_1 - \overline{X}_2$ is either too large or too small.

To find P, the probability of obtaining an $\overline{X}_1 - \overline{X}_2$ as unusual as -2.71 g/cc if actually $\mu_1 - \mu_2 = 0$, the z value corresponding to $\overline{X}_1 - \overline{X}_2 = -2.71$ is computed using the test statistic:

$$z = \frac{(\overline{X}_1 - \overline{X}_2) - (\mu_1 - \mu_2)}{\sigma\sqrt{1/n_1 + 1/n_2}}$$

or

$$z = \frac{-2.71 - 0}{0.3687} = -7.35$$

From Table A.2, the area to the left of $z = 3.99$ is 1.0000, so that the area to the right of 3.99, correct to four decimal places, is .0000, or in other words, $< .00005$. The area to the right of 7.35 is certainly $< .00005$. The probability of $\overline{X}_1 - \overline{X}_2$ being > 2.71 is $< .00005$, and the probability of $\overline{X}_1 - \overline{X}_2$ being smaller than -2.71 is $< .00005$. Summing the area in the two tails, $P < .0001$.

For any reasonable values of α, the conclusion is that there is a real difference in mean hemoglobin level between acyanotic and cyanotic children. In making this test, we are assuming that we have two independent simple random samples and that the computed z has a normal distribution.

7.2.2 Using a One-Sided Test

A one-sided test could also be used. Because acyanotic children are less severely handicapped than cyanotic children, we may believe that their mean hemoglobin level (μ_1) is lower than μ_2. The question then asked is "Is the mean hemoglobin level for acyanotic children lower than for cyanotic children?" In symbols, the question can be written as (is $\mu_1 < \mu_2$?). The null hypothesis is stated in the opposite direction, that is the mean level for acyanotic children is greater than or equal to that of cyanotic children or H_0: $\mu_1 \geq \mu_2$. If we are able to reject this null hypothesis, we can state that μ_1 is $< \mu_2$, and then from the value of P we know our chance of making an error.

Now a one-sided test is appropriate, and we calculate the chance of obtaining an $\overline{X}_1 - \overline{X}_2$ as small as -2.75 g/cc or smaller. From the results obtain previously, $P < .00005$, and the conclusion is that the hemoglobin level of acyanotic children is lower on the average than that of cyanotic children.

Note the difference in the conclusion between the two-sided and the one-sided tests. In the two-sided test, the question asked is whether μ_1 and μ_2 are different, and the conclusion is either that they are equal or that they are unequal; in the one-sided test the question asked is whether μ_1 is $< \mu_2$, and the conclusion is either that μ_1 is $< \mu_2$ or μ_1 is $\geq \mu_2$.

Sometimes the question asked is whether μ_1 is $> \mu_2$. Then, the rejection region in the right side of the distribution of $\overline{X}_1 - \overline{X}_2$ is used, and the conclusion is either that μ_1 is $> \mu_2$ or μ_1 is $\leq \mu_2$.

7.2.3 Testing for Equality of Means when σ Is Unknown

Usually, the variances of the two populations are unknown and must therefore be estimated from the samples. The test for determining if two population means are significantly different when the variances are unknown is one of the most commonly used statistical tests. In making this test, we assume that we have simple random samples chosen independently from two populations and that observations in each population are normally distributed with equal variances. The test is used either when we take samples from two distinct populations or when we sample a single population and randomly assign subjects to two treatment groups. This test has been shown to be insensitive to minor departures from the normal distribution.

To illustrate testing for equality of population means when the population variance is unknown, we will use the hemoglobin data for children with cyanotic and acyanotic congenital heart disease given in Table 7.2.

Just as in Section 6.5.3, when confidence intervals were found, we will use the pooled variance, s_p^2, as an estimate of σ^2, the population variance for each population. The pooled variance is computed as

$$s_p^2 = \frac{(n_1 - 1)s_1^2 + (n_2 - 1)s_2^2}{n_1 + n_2 - 2}$$

where s_1^2 is the variance from the first sample and s_2^2 is the variance from the second sample. From Table 7.2, we have $s_1^2 = 1.0167$, $s_2^2 = 1.9898$, $n_1 = 19$, and $n_2 = 12$. The computed pooled variance is

$$s_p^2 = \frac{18(1.0167) + 11(1.9898)}{19 + 12 - 2} = \frac{40.1884}{29} = 1.3858$$

Note that the pooled variance 1.3858 is between the numerical values of $s_1^2 = 1.0167$ and $s_2^2 = 1.9898$. It is somewhat closer to s_1^2 since $n_1 - 1$ is $> n_2 - 1$. To obtain the pooled standard deviation, s_p, we take the square root of 1.3858 and obtain $s_p = 1.1772$.

The standard deviation of the $\overline{X}_1 - \overline{X}_2$ distribution is

$$\sigma\sqrt{1/n_1 + 1/n_2}$$

It is estimated by

$$s_p\sqrt{1/n_1 + 1/n_2} = 1.1772\sqrt{1/19 + 1/12} = 1.1772\sqrt{0.13596} = 0.4341 \text{ g/cc}$$

The test statistic is

$$t = \frac{(\overline{X}_1 - \overline{X}_2) - (\mu_1 - \mu_2)}{s_p\sqrt{1/n_1 + 1/n_2}}$$

or

$$t = \frac{(\overline{X}_1 - \overline{X}_2) - 0}{s_p\sqrt{1/n_1 + 1/n_2}}$$

or for the results from Table 7.2,

$$t = \frac{13.03 - 15.74}{.4341} = \frac{-2.71}{.4341} = -6.24$$

Table A.3 is then consulted to find the chance of obtaining a t value at least as small as $t = -6.24$. The d.f. to be used in the table are

$$n_1 + n_2 - 2 = 19 + 12 - 2 = 29$$

(as usual the d.f. are the numbers used in the denominator of the estimate of the variance s_p^2). Looking at the row in Table A.3 that corresponds to d.f. = 29 we see that $t[.9995] = 3.659$, so that $> .9995$ of the t distribution lies below 3.659 and < 0.0005 lies above 3.659. The area below -3.659 is by symmetry also < 0.0005. Summing the area in both tails we have $P < .001$.

The null hypothesis of equal means is rejected; the conclusion is that the mean hemoglobin level for acyanotic children is different than for cyanotic children. The mean level for the acyanotic children is $\overline{X}_1 = 13.03$ and that of the cyanotic children $\overline{X}_2 = 15.74$. As before, $\overline{X}_1 - \overline{X}_2 = -2.71$, so if we reject the null hypothesis of equal means we can also say that acyanotic children have lower mean hemoglobin levels with $P < .001$.

The *t test for independent groups* (unpaired) just described is one of the most commonly used statistical tests. It is widely available both in statistical programs and even in some spreadsheet programs. In most programs, the results are given for the two-sided test. Most programs print out t and the actual level of P instead of simply saying it is less than a given value. The user then compares the computed P value to a prechosen α level or simply reports the P value.

In most statistical programs, one variable is used to list the hemoglobin levels for both groups. A second variable, often called a grouping or class variable, tells the computer program which group the observation comes from. In the hemoglobin example, the class or grouping variable could be assigned numerical values of 1 or 2 to indicate whether the child was cyanotic or acyanotic.

7.3 TESTING FOR EQUALITY OF MEANS: PAIRED DATA

Often in comparing two populations, the sample data occur in pairs. The researchers studying congenital heart disease who wish to compare the development of cyanotic children with normal children might gather data on age at first word for two independent samples: one of cyanotic children and the other of well children. Then, we could test the null hypothesis that the two population means are equal as in Section 7.2.3. Instead, however, available data (see Table 7.3) might consist of age of first word for 10 pairs of children, with each pair consisting of a cyanotic child and a normal sibling. Because children in the same family tend to be alike in many respects, the data cannot be considered as two independent samples, and should be treated as a sample of pairs.

Instead of analyzing the ages directly, the 10 differences are formed as in Table 7.3, and then are treated as a single sample from a population of differences. If d represents a difference, and μ_d the mean of the population of differences, then if the cyanotic children and siblings learn to talk at the same time, on the average, $\mu_d = 0$.

Table 7.3. Age at First Word in Months for 10 Children with Cyanotic Heart Disease and Their 10 Siblings

Pair Number	Cyanotic X_1	Sibling X_2	Difference d
1	11.8	9.8	2.0
2	20.8	16.5	4.3
3	14.5	14.5	0.0
4	9.5	15.2	−5.7
5	13.5	11.8	1.7
6	22.6	12.2	10.4
7	11.1	15.2	−4.1
8	14.9	15.6	−0.7
9	16.5	17.2	−0.7
10	16.5	10.5	6.0
			$\overline{d} = 1.32$
			$s_d^2 = 22.488$

Either a one-sided or a two-side test can be performed. If the question to be answered by the test is whether cyanotic children learn to talk later than siblings, then a one-sided test is appropriate. Rephrased in terms of μ_d, the question becomes: Is $\mu_d > 0$? A one-sided, upper tail test is appropriate. The null hypothesis is H_0: $\mu_d \leq 0$ so if it is rejected we can say that $\mu_d > 0$. In Table 7.3, the mean of the 10 differences is calculated to be $\overline{d} = 1.32$ months; the variance of the differences is $s_d^2 = \sum(d - \overline{d})^2/(n-1) = 22.488$ and their standard deviation is $s_d = 4.74$ months. The variance of the population of sample means is estimated by

$$s_{\overline{d}}^2 = \frac{s_d^2}{n} = \frac{22.488}{10} = 2.2488$$

and by taking the square root of 2.2488 we have 1.50 months as an estimate of the standard deviation $s_{\overline{d}}$ of the $\overline{d}$ population. Then,

$$t = \frac{\overline{d} - 0}{s_{\overline{d}}} = \frac{1.32 - 0}{1.50} = 0.880$$

With $\alpha = .05$ and d.f. equal to $n - 1 = 9$, a $t[.95]$ value of 1.833 would be necessary to reject the null hypothesis (see Table A.3). With $t = 0.880$ far less than 1.833, we cannot reject the null hypothesis; the mean age of first word may be the same for cyanotic children as for their normal siblings. Alternatively, the P value for the test can be obtained; from Table A.3, 75% of the area is below 0.703; 90% lies below 1.382. The proportion of t values below 0.880 is between 75 and 90%, and P, the proportion above 0.880 is between 10 and 25% or $P < 0.25$.

In making this test, we assumed that the d's are a simple random sample from a normal distribution. If we are unsure whether the differences being normally distributed, then either a normal probability plot, a histogram, or a box plot should be graphed. The differences should be at least approximately normally distributed.

When a computer program is used, the observations for the cyanotic children and for their siblings are usually entered as separate variables. In some programs, the user first has the program make a new variable, which is the difference between the ages for the cyanotic children and their siblings, and then treats the problem as if it were a single sample of observations and performs a t test for a single sample. In other programs, the user tells the program which variable is the first variable and which is the second, and the program gives the result of the paired t test. Usually the P value for a two-sided test is given.

7.4 CONCEPTS USED IN STATISTICAL TESTING

In this section, we discuss the decision to accept or to reject a null hypotheses. The two types of error are defined and explained.

7.4.1 Decision to Accept or Reject

After testing a null hypothesis, one of two decisions is made; either H_0 is accepted or it is rejected. If we use a significance level $\alpha = .05$, and reject H_0: $\mu = 12.0$, then we feel reasonably sure that μ is not 12.0 because in the long run, in repeated experimentation, if it actually is 12.0 the mistake of rejecting $\mu = 12.0$ will occur only 5% of the time.

On the other hand, if, with $\alpha = .05$, H_0 is accepted, it should be emphasized that we should not say that μ is 12.0, instead we say that μ *may be* 12.0.

In a sense, the decision to reject H_0 is a more satisfactory decision than is the decision to accept H_0, since if we reject H_0 we are reasonably sure that μ is not 12.0, whereas if we accept H_0, we simply conclude that μ may be 12.0. The statistical test is better adapted to "disproving," than to "proving". This is not surprising. If we find that the available facts do not fit a certain hypothesis, we discard the hypothesis. If, on the other hand, the available facts seem to fit the hypothesis, then we do not know whether the hypothesis is correct or whether some other explanation might be even better.

After we test and accept H_0, we have just a little more faith in H_0 than we had before making the test. If a hypothesis has been able to stand up under many attempts to disprove it, then we begin to believe that it is a correct hypothesis, or at least nearly correct.

7.4.2 Two Kinds of Error

In a statistical test, two types of mistake can occur in making a decision. We may reject H_0 when it is actually true, or we may accept it when it is actually false. The two kinds of error are analogous to the two kinds of mistake a jury can make. The jury may make the mistake of deciding that the accused is guilty when actually innocent, or the jury may make the mistake of deciding that the accused is innocent when actually guilty.

The first type of error (rejecting H_0 when it is really true) is called a Type I error and the chance of making a Type I error is called α. If we reject the null hypothesis and if α is selected to be .01, then if H_0 is true, we have a 1% chance of making an error and deciding that H_0 is false.

Earlier in this chapter we noted that often, though not always, α is set at .05. In certain types of biomedical studies, the consequences of making a Type I error are worse than in others. For example, suppose the medical investigator is testing a new vaccine for a disease for which no vaccine exists at present. Normally, either the new vaccine or a placebo vaccine is randomly assigned to volunteers. The null hypothesis is that both the new vaccine and the placebo vaccine are equally protective against the disease. If we reject the null hypothesis and conclude that the vaccine has been successful in preventing the disease when actually it was not successful, then the consequences are serious. A worthless vaccine might be given to millions of people. Any vaccine has some risks attached to using it, so many people could be put at risk. Also, a great deal of time and money would have been wasted. In this case, an α of .05 does not seem small enough.

It may then seem that a Type I error should always be very small. But this is not true because of a second type of error.

The second type of error, accepting H_0 when it is really false, is called a Type II error. The probability of making this type of error is called β. The value of β depends on the numerical value of the unknown true population mean, and the size of β is larger if the true population mean is close to the null hypothesis mean than if they are far apart.

In some texts and statistical programs a Type I error is called an α-error and Type II error is called a β-error.

Decisions and outcomes are illustrated in Table 7.4. Since we either accept or reject the null hypothesis, we know after making our decision which type of error we may have made. Ideally, we would like our chance of making an error very small.

If our decision is to reject the H_0, then we either made a correct decision or a Type I error. Note that we set the size of α. The chances of our making a correct decision is $1 - \alpha$.

If our decision is to accept the null hypothesis, then we have made either a correct decision or a Type II error. We do not set the size of β before we make a statistical test and do not know its size after we accept the null hypothesis.

Table 7.4. Decisions and Outcomes

	Accept H_0	Reject H_0
H_0 true	Correct decision	Type I error
H_0 false	Type II error	Correct decision

We can use a desired size of β in estimating the needed sample size as seen in Section 7.5.

7.4.3 An Illustration of β

To illustrate the Type II error, we will examine a simple hypothetical example. We have taken a simple random sample of $n = 4$ observations and we know that $\sigma = 1$ so that $\sigma/\sqrt{4} = 0.5$. The H_0: $\mu = 0$ is to be tested and we set $\alpha = .05$. First, we will find the values of $\overline{X}$ that separate the rejection from the acceptance region. The value of $\overline{X}$ corresponding to $z = +1.96$ is obtained by solving

$$1.96 = \frac{\overline{X} - 0}{0.5}$$

for $\overline{X}$. Thus, $\overline{X} = 1.96(0.5) = 0.98$ corresponds to $z = 1.96$ and similarly $\overline{X} = -0.98$ corresponds to $z = -1.96$. Any value of $\overline{X}$ between $-.98$ and $+.98$ would cause us to accept the null hypothesis that $\mu = 0$; thus β equals the probability that $\overline{X}$ lies between $-.98$ and $+.98$ given that we have actually taken the sample from a population whose μ has a true value not equal to $\mu = 0$. Let us suppose that the true value of μ is 0.6.

Figure 7.4 (*a*) represents the normal distribution of the $\overline{X}$'s about $\mu = 0$ with the acceptance region in the center and the rejection regions in both tails. Figure 7.4 (*b*) represents the normal distribution of $\overline{X}$'s about the true value of $\mu = 0.6$. The two normal curves are lined up vertically so that zero on the upper curve is directly above zero on the lower curve. In Figure 7.4(*b*), the shaded area is β and the unshaded area is $1 - \beta$, since the total area under the curve is 1. We can see that β is quite large.

We can imagine moving the lower Figure 7.4(*b*) to the right. In this case, less of the area in Figure 7.4(*b*) would be within the acceptance region of the upper Figure 7.4(*a*) and we have reduced the chance of making a Type II error. Also, if we move the lower figure to the left, say make the true $\mu = 0.1$, then we increase our chance of making a Type II error, since more of the area of the lower curve lies within the acceptance area of the upper curve but the null hypothesis is not true. If we moved the lower figure very far to the left, then the chance of making a Type II error is reduced.

There are several general principles that can be drawn from examining figures similar to Figure 7.4(*a*) and (*b*) for two-sided tests.

1. For a given size of α, σ, and n, the farther the hypothesized μ is from the actual μ, the smaller β will be if the null hypothesis is accepted. If we accept the null hypothesis, we are most apt to make a Type II error if the hypothesized mean and the true mean are close together.
2. If we set the value of α very small, then the acceptance region is larger than it would otherwise be. We are then more likely to accept the null hypothesis if it is not true; we will have increased our chance of making a Type II error. Refer to Figure 7.4(*a*) and imagine the rejection region being made smaller in both tails, thus increasing the area of β in Figure 7.4(*b*).

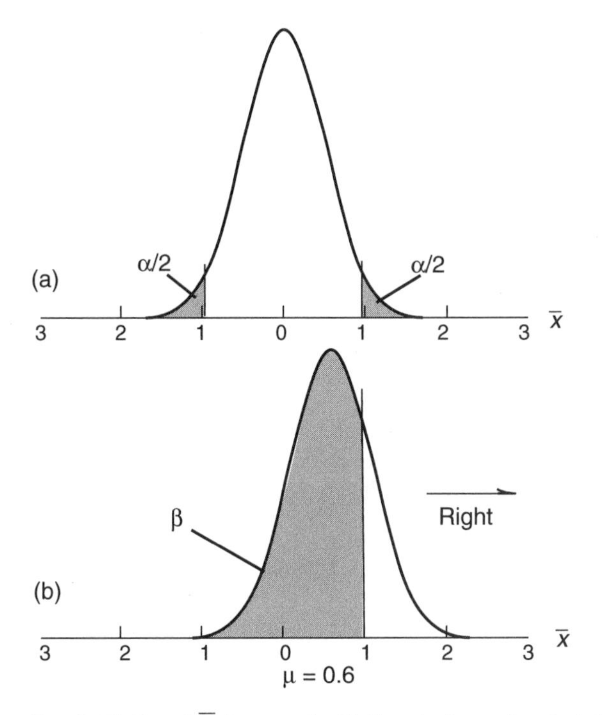

Fig. 7.4. (*a*) Sampling distribution of $\overline{X}$ for $\mu = 0$ with the rejection region shaded. (*b*) Sampling distribution of $\overline{X}$ when true $\mu = 0.6$ with the shaded area β.

3. If we increase the size of n, we will decrease the $\sigma_{\overline{X}}$ (standard error of the mean) and both normal curves will get narrower and higher (imagine them being squeezed together from both sides). Thus, the overlap between the two normal curves will be less. In this case, for a given preset value of α and the same mean values, we will be less apt to make a Type II error since the curves will not overlap so much.

Since β is the chance of making a Type II error, then $1 - \beta$ is the chance of not making a Type II error and is called the *power* of the test. When making a test, we decide how small to make α and try to take a sufficient sample size so that the power is high.

7.5 SAMPLE SIZE

When planning a study, we want our sample size to be large enough so that we can reach correct conclusions but we do not want to take more observations than necessary.

We will illustrate the computation of the sample size first for a one-sided test of the mean for a single sample. The formula for sample size is derived by solving n for a given value of α, β, σ, and $\mu_0 - \mu$ (the difference between the hypothesized mean and the actual mean). For a one-sided test, the needed sample size is

$$n = \frac{\sigma^2(z[1-\alpha] + z[1-\beta])^2}{(\mu - \mu_0)^2}$$

Suppose we wish to compute the needed sample size for the study reported in Table 7.1. Here we assumed that $\sigma = 1.75$ and $\mu_0 = 12$ months. Often, α is chosen to be .05 and $\beta = .20$ or $1 - \beta$ is .80. Other values can be taken based on the seriousness of making a Type I or Type II error. From Table A.2, we obtain $z[.95] = 1.65$ and $z[.80] = 0.84$. The value of μ is often chosen based on what size difference from μ_0 it is important to detect. In other words, if the age at walking was only 1 week less for the children who have acyanotic congenital heart disease most investigators might not think that was any concern. Here, we will arbitrarily decide that we will want to reject the null hypothesis if the difference is at least 1 month (if μ is < 11 months). We would then enter these values into the sample size formula and obtain

$$n = \frac{1.75^2(1.65 + 0.84)^2}{(11 - 12)^2}$$

or

$$n = \frac{(3.0625)(6.2)}{1} = 18.98$$

We will round up 18.98 to 19 and decide that our sample size should be 19.

In examining the formula for sample size, we note that the needed sample size will be large if we have a large σ, if we want a small chance of making a Type I and Type II error, and if the difference we want to find between the actual mean and the hypothesized mean is small. In other words, the results reflect what was seen in Figure 7.4.

For a two-sided test, the results are approximate. The appropriate formula is the same as for the one-sided test except we will use $z[1 - \alpha/2]$ in place of $z[1 - \alpha]$. Note that in a two-sided test, each rejection region has a size of $\alpha/2$. The approximate result is

$$n = \frac{\sigma^2(z[1-\alpha/2] + z[1-\beta])}{(\mu - \mu_0)^2}$$

Thus, for $\alpha = .05$, we would replace 1.65 by 1.96 and the calculations would proceed in the same manner. The required sample size would be larger than for a one-sided test.

For a one-sided test of means from two *independent* samples, the formula for the sample size is quite similar to the one given for a single sample. Here, we will assume that the estimated sample size in the two groups is equal, $n_1 = n_2 = n$ and again that the σ in both groups is the same. As shown earlier in Chapter 6, the variance of $\overline{X}_1 - \overline{X}_2 = 2\sigma^2/n$, so that we will replace σ^2 with $2\sigma^2$. We will also replace $\mu - \mu_0$

with $\mu_1 - \mu_2$ since now the null hypothesis is stated in terms of the means of the two populations. For a one sided test, the sample size can be obtained from

$$n = \frac{2\sigma^2(z[1-\alpha] + z[1-\beta])^2}{(\mu_1 - \mu_2)^2}$$

and for a two-sided test the approximate formula for n is given by

$$n = \frac{2\sigma^2(z[1-\alpha/2] + z[1-\beta])^2}{(\mu_1 - \mu_2)^2}$$

Since the two-sided test for testing the H_0: $\mu_1 = \mu_2$ is such a widely used test, the formula for obtaining the needed sample size for this case is also widely used. From examining the formula, it is obvious that we have to choose a value for α and β. As stated previously, $\alpha = .05$ and $\beta = .20$ are commonly used values; for these values $z[1-\alpha/2] = 1.96$ and $z[1-\beta] = 0.84$ or more accurately 0.842. Other values of α and β can be chosen, depending on the seriousness of making Type I and Type II errors.

We next decide what *difference* between μ_1 and μ_2 we wish to detect. Note that the actual values of μ_1 and μ_2 are not needed. For example, if the outcome variable was systolic blood pressure, a difference as small as 2 mmHg would probably not be of clinical significance, so something larger would be chosen. The size should depend on what is an important difference to be able to detect. A clinician might say 5 mmHg is a sizable difference to detect so 5 would be used in the formula in place of $\mu_1 - \mu_2$. An estimate of σ is also needed and the clinician may decide from past experience that $\sigma = 15$ mmHg. Then, the desired sample size is

$$n = \frac{2(15)^2(1.96 + 0.84)^2}{(5)^2} = \frac{(450)(7.84)}{25} = 141.12$$

so that $n = 142$ are needed in each group.

If σ is unknown so that a t test will be performed using the pooled estimate s_p the above formula is still used. Often we will make the best estimate of σ we can either by looking up similar studies in the literature or from past results in the same hospital. Alternatively, the approximate method for estimating σ from the range given in Section 4.2.2 can be used. If it is not possible to make an estimate of σ, then the only way the sample size formula for n can be used is to decide what value of $(\mu_1 - \mu_2)/\sigma$ the user wishes to detect. For example, in the systolic blood pressure example, the choice of values for σ and $\mu_1 - \mu_2$ results in $\frac{5}{15} = \frac{1}{3} = 0.33$ as the difference between the means relative to the standard deviation that we wish to detect. Values between $\frac{1}{5}$ and 1 are often chosen. The formula for n uses the inverse of $(\mu_1 - \mu_2)/\sigma$, that is, $\sigma/(\mu_1 - \mu_2) = 3$ is used for the systolic blood example, and the resulting formula becomes

$$n = (\sigma/(\mu_1 - \mu_2))^2 2(z[1-\alpha/2] + z[1-\beta])^2$$

or

$$n = 3^2 2(1.96 + 0.84)^2 = 141.2$$

which is rounded up to 142. If the computed sample size is small and a t test with s_p will be used later in testing, some investigators will increase the estimated sample size by several observations.

For paired observations, there are two likely choices for estimating the sample size. If the user can estimate σ_d, then the formula for a single sample can be used with σ replaced by σ_d and with μ_d replacing $\mu - \mu_0$. Usually, a good estimate of the σ_d is not available. Commonly, the formula for two independent groups is used with the thought that at least a sufficiently large estimate of n will be obtained.

Special statistical programs are available that provide sample sizes for a wide range of statistical tests. Sample sizes for numerous tests can also be found in Kraemer and Thiemann [1987].

7.6 CONFIDENCE INTERVALS VERSUS TESTS

In the examples discussed in this chapter and in Chapter 6, we analyzed the data using either confidence intervals or tests of hypotheses. When are confidence intervals usually used and when are tests used? What are the relative advantages of the two methods?

Most investigators run tests on the variables that are used to evaluate the success of medical treatments and report P values. For example, if a new treatment has been compared to a standard treatment, the readers of the results will want to know what the chance of making a mistake is if the new treatment is adopted. Suppose the variable being used to evaluate two treatments is length of stay in the hospital following treatment. Then, for that variable an investigator will likely perform a statistical test.

For many studies, the investigators are not testing treatments, but instead are trying to determine if two groups are different. This was the case in the example in this chapter when the hemoglobin levels of cyanotic and acyanotic children were analyzed. Here, either confidence intervals or tests could be done but probably a stronger case could be made for confidence intervals. In general, it is sensible to run tests when investigators have a rationale for the size of the P value that would lead them to reject the null hypothesis and when users of the results want to know their chance of making a Type I error. For further discussion of confidence intervals and tests of hypotheses, see Fisher and van Belle [1993].

Confidence intervals are often used when describing the patients at their entrance to the study. They are useful when the main interest is in determining how different the population means are. Since confidence intervals provide limits that have a known chance of including the difference in two population means they directly answer this question. The questions asked in many studies by epidemiologists are best answered with confidence intervals. See Rothman [1986].

In practice, tests are often reported because statistical programs tend to provide output on test results. Programs do provide the means and standard deviations from which confidence intervals can be easily computed but they do not always display the actual confidence intervals.

7.7 CORRECTING FOR MULTIPLE TESTING OR CONFIDENCE INTERVALS

When we make multiple tests from a single data set, we know that with each test we reject, we have an α chance of making a Type I error. This leaves us with the uncomfortable feeling that if we make enough tests our chance that at least one will be significant is $> \alpha$. For example, if we roll a die our chances of getting a 1 is only 1 in 6. But if we roll the die numerous times our chance of getting a 1 at some time increases.

Suppose we know in advance that we will make m tests and perform two-sided tests. If we want to have an overall chance of making an Type I error be $\leq \alpha$, then we compare the computed t values we obtain from our computations or the computer output with $t[1-\alpha/2m]$ using the usual d.f. instead of $t[1-\alpha/2]$. For example, suppose we know we will make $m = 4$ tests and want $\alpha = .05$. Each test has 20 d.f. In this case, $2m = 8$. From a computer program, we obtain $t[1-.05/8] = t[.9938] = 2.748$. We can see from Table A.3 that $t[.9938]$ with 20 d.f. lies between 2.528 and 2.845, so the t value of 2.748 obtained from a statistical program seems reasonable. Suppose our four computed t values were 1.54, 2.50, 2.95, and 3.01. If we correct for multiple testing, only the t tests that had values of 2.95 and 3.01 would be significant since they exceed 2.748. Without correcting for multiple testing, we would also reject the second test with a t value of 2.50 (see Table A.3). To correct for multiple testing using one-sided tests, we compare the computed t's with tabled $t[1 - \alpha/m]$ instead of tabled $t[1 - \alpha]$ values.

This method of correcting for multiple testing is perfectly general and works for any type of test. It is called the Bonferroni correction for multiple testing.

The Bonferroni correction also works for confidence intervals. Here, for a two sided interval with σ known, we use $z[1 - \alpha/2m]$ instead of $z[1 - \alpha/2]$ when we compute the m confidence intervals. For σ unknown, we use $t[1 - \alpha/2m]$ instead of $t[1 - \alpha/2]$ in computing the intervals. When using t, it may be difficult to find tabled values close to those you want. The simplest thing to do is to use a computer program that gives t values for any number between 0 and 1. As a rough approximation, linear interpolation given in Section 5.2.2 may be used. A better approximation is given in Fleiss [1986].

7.8 REPORTING THE RESULTS

In typical biomedical examples involving patients, numerous variables are measured. A few of these variables relate to the main purpose of the study and are called outcome variables. Here, tests of hypotheses are commonly reported. But other variables are collected that describe the individual patients such as age, gender, seriousness of medical condition, results from a wide range of medical tests, attitudes, and previous treatment. These data are taken so that the patients being treated can be adequately described and to see if the patients in the different treatment groups were similar prior to treatment. In experiments using animals and in laboratory experiments, usually fewer variables are measured. In reporting the results, we have to decide what is of interest to the reader.

When we report the results of tests of hypotheses, we should include not only the P value but also other information from the sample that will aid in interpreting the results. Simply reporting the P value is not sufficient. It is common when reporting the results of H_0: $\mu_1 = \mu_2$ to also report the two sample means so the reader can see their actual values. Usually, either the standard deviations or the standard errors of the means are also included. The standard deviation is given if we want the reader to have a measure of the variation of the observations and the standard error of the mean is given if it is more important to know the variation of the mean value. The standard deviation is sometimes easier to interpret than the standard error of the mean if the sample size in the two groups are unequal since its size does not depend on the sample size.

Often tables of means and standard deviations are given for variables that simply describe the patients in the study. This information is often given in a graphical form with mean comparison charts (see Section 4.6) or other types of graphs.

If the shape of the distribution is of interest, then using histograms, box plots, or other available options are useful. These plots can be obtain directly from the statistical programs and many readers find them easier to interpret than tables of statistics.

PROBLEMS

In the following problems, use the steps listed in Section 7.1.3 as a guideline when performing tests.

7.1 Using each of the samples from the cholesterol population in Problem 2.1, test H_0: $\mu = 279.5$ (three separate tests, one for each sample). Assume that you know $\sigma = 45.3$ mg/100 mL.

7.2 Of all the tests made in Problem 7.1 by the class, how many rejected H_0? About how many would you expect to reject H_0? Explain why the number of rejections might differ from the number you expect.

7.3 In Problem 6.5, make a test to determine whether pressure increases mean bleeding time or not, Use $\alpha = .05$.

7.4 In Problem 6.4, test whether male rats are heavier, on the average, than female rats. Use $\alpha = .05$.

7.5 In Problem 6.8, make a test to determine whether isoniazid treatment for leprosy was effective. Use $\alpha = .05$.

7.6 In Problem 6.7, test to determine (a) Whether oral and rectal temperatures are the same. (b) Whether mean oral temperature is 37°C. Use $\alpha = .05$ in both (a) and (b).

7.7 The following data are from the U.S. Bureau of Census. They have been listed by New England and Pacific states according to the Census classification system. The first column of data gives the number of patient care physicians per 10,000 state population in 1995. The second column is the number of hospital beds per 1000 civilian population in 1994. The third column is the percent of persons without health care coverage in 1995. The fourth column is the percent of the population that are $\geq$ 65 years old.

Location	Physicians	Beds	No Coverage	$\geq$ 65
New England				
Maine	18.2	3.4	13.5	12.0
New Hampshire	19.8	2.9	10.0	12.0
Vermont	24.2	3.3	13.2	12.1
Massachusetts	33.2	3.3	11.1	14.1
Connecticut	29.5	2.8	8.8	14.3
Pacific				
Washington	20.2	2.8	12.4	11.6
Oregon	19.5	2.3	12.5	13.4
California	21.7	2.5	20.6	11.0
Alaska	14.2	2.2	12.5	5.2
Hawaii	22.8	2.6	8.8	12.9

If we were to perform a *t* test for the each of the four variables listed above, testing in each case that the mean for the New England equals the mean for the Pacific region, what assumption do we clearly not meet? If we wished to correct for multiple testing what would we do?

7.8 Investigators are studying two drugs used to improve the FEV_1 readings (a measure of lung function) in asthmatic patients. Assume that $\sigma = 0.8$ and the difference in means they wish to detect is 0.25. If they want to do a two-sided test of equal population means with $\alpha = .05$ and $\beta = .20$, what sample size should they use?

7.9 Test the null hypothesis of equal means for the data given in Problem 6.6 using $\alpha = .01$.

REFERENCES

Fisher, L. D. and van Belle, G. [1993]. *Biostatistics: A Methodology for the Health Sciences,* New York: Wiley-Interscience, 106–115.

Fleiss, J. L. [1986]. *The Design and Analysis of Clinical Experiments,* New York: Wiley-Interscience, 104–105.

Kraemer, H. C. and Thiemann, S. [1987]. *How Many Subjects?*, Newbury Park, CA: Sage.

Rothman, K. J. [1986]. *Modern Epidemiology,* Boston, MA: Little, Brown and Company, 120–121.

CHAPTER EIGHT

Categorical Data: Proportions

In previous chapters, the data considered were in the form of continuous measurements; for each individual in a sample some characteristic was measured: height, weight, blood cholesterol, gain in weight, temperature, or duration of illness. This chapter and Chapter 9 are both concerned with a somewhat different type of data: data consisting merely of counts. For example, the number of survey respondents who are male or female may be counted. Here the variable is gender and there are two categories: male and female. The proportion of males would be the count of the number of males divided by the total number of males and females. Some variables such as race or religion are commonly classified into more than two categories.

According to Stevens' system for classifying data, the data could be nominal or ordinal. Sometimes biomedical researchers also like to group continuous (interval or ratio) data into separate categories. For example, the ages of a sample of adults could be split into three categories, young, middle aged, and old.

In this chapter, we give methods of handling categorical data for populations whose individuals fall into just two categories. For example, for young patients who underwent an operation for cleft palate, the two outcomes of the operation might be no complications (success), and one or more complications (failure). This type of data is essentially "yes" or "no", success or failure. Categorical data with more than two categories will be discussed in Section 9.4.

In Section 8.1, methods for calculating the population mean and variance from a single population are presented. Here, the formulas for the population parameters are given before those for the sample. The relationship between the sample statistics and the population parameters is presented in Section 8.2. The binomial distribution is introduced, and we explain that in this text only the normal approximation to the binomial distribution is covered. Section 8.3 covers the use of the normal approximation. Confidence limits for a single population proportion are given in Section 8.4 and for the difference between two proportions in Section 8.5. Tests of hypotheses are discussed in Section 8.6 and the needed sample size for testing two proportions is given in Section 8.7. When either confidence intervals or tests of hypothesis are given, we will assume that simple random samples have been taken. Section 8.8 covers data entry of categorical data and typical output from statistical programs.

8.1 SINGLE POPULATION PROPORTION

Here, we consider the young patients who underwent the cleft palate operation to be the entire population; we have 5 who had a complication (failures) and 15 who had no complications (successes). To make it simpler to count the number of successes and failures, we shall code a success as a 1 and a failure as a 0. We then have 15 ones and 5 zeros and $N = 20$ is the total number of observations in the population. The data are usually reported in terms of the proportion of *successes* (the number of successes over the total number of observations in the population). Since in this example we are reporting the results in terms of successes, we coded the successes as a 1 and the failures as a 0. For our population of young patients, this proportion of successes is $15/20 = .75$ and .75 is equivalent to the mean of the population since we have divided the sum of the numerical values of the observations by N. That is, if we add the 15 ones and 5 zeros we get 15, and 15 divided by 20 is .75. Similarly, if we count the number of successes and divide by the number of observations we get 15 over 20 or .75. The population mean is called π. In this population, the population mean is $\pi = .75$. This is also the population proportion of successes.

The proportion of successes and the proportion of failures must add to one since those are the only two possible outcomes. Hence, the proportion of failures is $1 - \pi$.

8.1.1 Graphical Displays of Proportions

Graphically, this type of data is commonly displayed as pie charts such as that given in Figure 8.1 or bar charts as shown in Figure 8.2. Pie charts are readily interpretable when there are only two outcomes; they are not recommended when the number of categories is large (see Cleveland [1985]).

In Figure 8.2, it can be seen that 15 of the patients had a successful operation. Bar charts have the advantage that they can be used with any number of categories. In pie or bar charts either counts, proportions, or percents can be displayed. Figure 8.2 displays the outcome as counts; there were 15 successes and 5 failures.

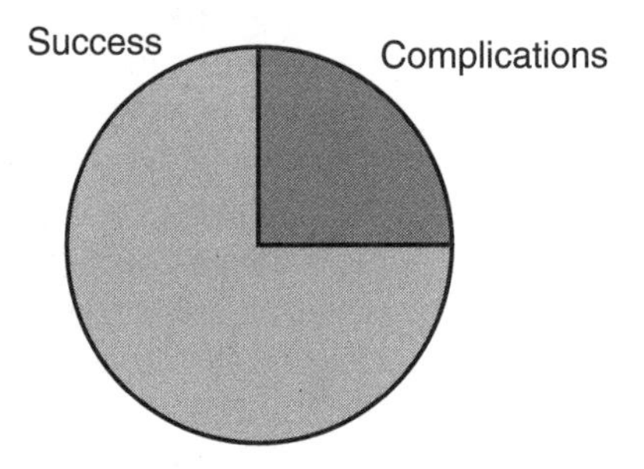

Fig. 8.1. Pie chart of proportion of successes and complications.

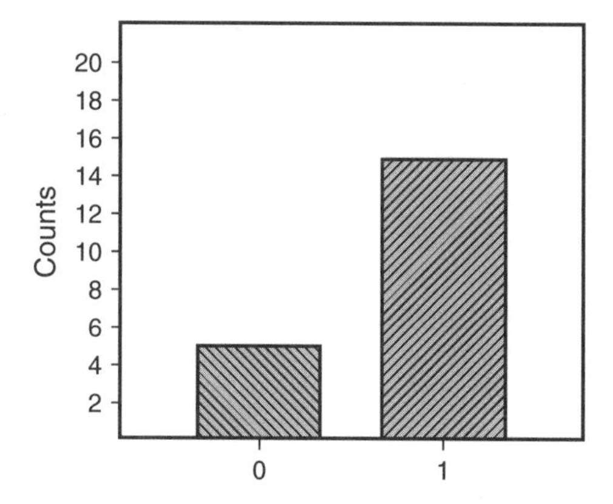

Fig. 8.2. Bar chart of counts of complications and successes.

Table 8.1. Computation of Population Variance for Proportions

X	f	fX	fX^2
1	$N\pi = 15$	$N\pi = 15$	$N\pi = 15$
0	$N(1 - \pi) = 5$	0	0
Sum	$N = 20$	$N\pi = 15$	$N\pi = 15$

8.1.2 Calculating the Population Variance

We now return to the population of 20 patients. We can calculate the population mean and variance from a frequency table using $\pi = \sum fX/N$ and the $\sigma^2 = [\sum fX^2 - N(\overline{X})^2]/N$ (see Section 4.3). Note that N is used instead of $n - 1$ in calculating the population variance. The number or frequency of successes is $N\pi = 15$ and the number of failures is $N(1 - \pi) = 5$. The results are shown in Table 8.1. In Table 8.1, f denotes frequency. Both the numerical values and the parameter for the population mean are included. Since the values of X are either 0 or 1, the table is very simple. Note also that when $X = 1$ that numerically $fX = fX^2$, since 1^2 is still 1.

We first compute the population mean or π. With the usual formula for the mean and the results given in the sum row, we have $\pi = \sum fX/N = 15/20 = .75$. Substituting the symbols for the population mean and the number of observations in the population from Table 8.1, we have $\pi = (\pi N)/N$ or simply π.

The variance is $(15 - (20)(.75)^2)/20 = (15 - 11.25)/20 = .1875$. In symbols we have

$$\sigma^2 = \frac{N\pi - N(\pi)^2}{N} = \frac{N(\pi - \pi^2)}{N} = \pi(1 - \pi)$$

as the population variance for the *observations*. We can compute the population variance simply as $.75(1 - .75) = .1875$.

The number π is a population parameter, just as are μ and σ; p is the corresponding *sample statistic*. The parameter μ and its estimate $\overline{X}$ were studied, in Section 4.4, by computing a large number of $\overline{X}$'s. This led to the study of the distribution of $\overline{X}$'s. Then, in Chapter 6 confidence intervals were constructed for μ (and also for $\mu_1 - \mu_2$, in the case of two populations). Finally, in Chapter 7 tests of hypotheses were made concerning μ (and concerning $\mu_1 - \mu_2$).

In this chapter, the distribution of p will be considered, and the same confidence intervals and tests of significance will be developed for π as were developed earlier for μ. This is done somewhat more briefly than with the measurement data, since the problems and the methods are analogous.

8.2 SAMPLES FROM CATEGORICAL DATA

In the problem set with continuous data, a population of blood cholesterol measurements was used and a large number of samples were drawn from it. The sample means were computed for each sample, and then these sample means were arranged in a frequency distribution in an attempt to study the distribution of all possible sample means. In this way, some of the rules that have been proved mathematically concerning the distribution of sample means were illustrated.

A box containing a large number of beads of two colors can be used to illustrate the process of drawing samples from a two-valued yes-or-no type of categorical data population. If some of the beads are white and the remainder are red, the white beads can represent the adults who have not had measles, and the red beads the adults who have. Or the white beads can be the patients for whom a new treatment is successful, and the red beads those for whom it is not. A large number of samples of some specified size (say 5) can be drawn from the population of beads, and for each sample the white beads can be counted and the sample proportion of white beads recorded.

It can be shown mathematically that if a very large number of samples of size 5 were drawn (with replacement) from a bead box, and if the proportion of white beads in the box is $\pi = .2$, the distribution of sample proportions occurring would be very close to that given in Table 8.2. If we kept sampling on and on from the bead box, then in the long run 40,960 out of every 100,000 samples would give $p = .2$ (the correct value), so that we obtain the correct value about 41% of the time. About 33% of the time (32,768 out of 100,000), we get $p = .0$; 20% of the time (20,480 out of 100,000) we get $p = .4$; about 5% of the time (5120 out of 100,000) we get $p = .6$; $< 1\%$ of the time we get $p = .8$ (640 out of 100,000); and it is quite unusual to get $p = 1$ (32 out of 100,000).

Frequency tables like Table 8.2 can be computed (and are available) for various values of π, the population proportion, and of n, the sample size. These are called *binomial distributions* and are computed using a mathematical formula. The mathematical formula for the binomial distribution is beyond the scope of this book, but Figure 8.3 gives the appearance of the distribution for $\pi = .2$ and $n = 5$, $\pi = .5$ and $n = 5$, $\pi = .2$ and $n = 10$, and $\pi = .5$ and $n = 10$. The distributions are seen to be clustered more closely around π for $n = 10$ than for $n = 5$. For an explanation of the

Table 8.2. Frequency Distribution of Sample Proportion from Population with $\pi = .2$ and $n = 5$

p	Frequency	Relative Frequency	fp	fp^2
.0	32,768	0.328	0	0.0
.2	40,960	0.410	8,192	1,638.4
.4	20,480	0.205	8,192	3,276.8
.6	5,120	0.051	3,072	1,843.2
.8	640	0.006	512	409.6
1.0	32	0.000	32	32.0
$\sum$	100,000	1.00	20,000	7,200.0

binomial distribution see Dixon and Massey [1983], Fisher and van Belle [1993], or Mould [1995]. From Table 8.2, the mean of the sample proportions can be computed as

$$\mu_p = \frac{20,000}{100,000} = .20$$

and the variance of the sample proportions is

$$\sigma_p^2 = \frac{7200.0 - 100,000(.20)^2}{100,000} = .032 = \frac{\pi(1-\pi)}{n}$$

Two statements can be proved mathematically about the distribution of all possible sample proportions; here they will be accepted without proof, and will be verified in the case of $\pi = .2$, $n = 5$.

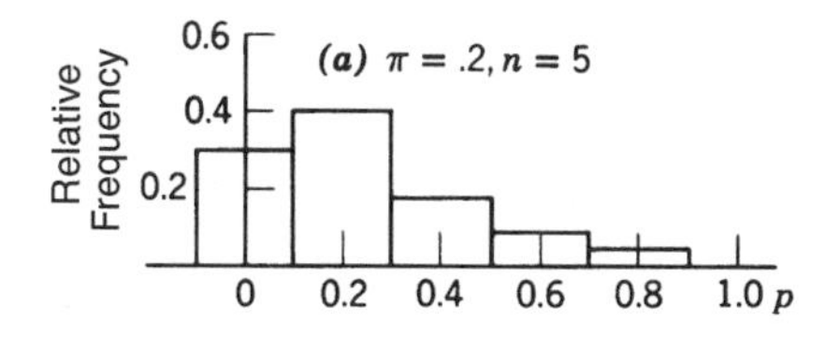

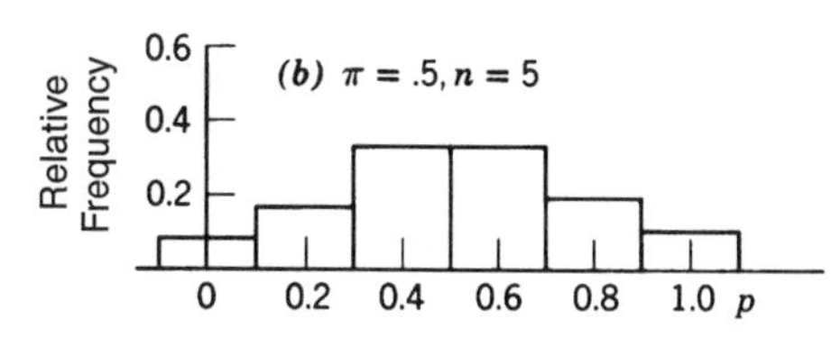

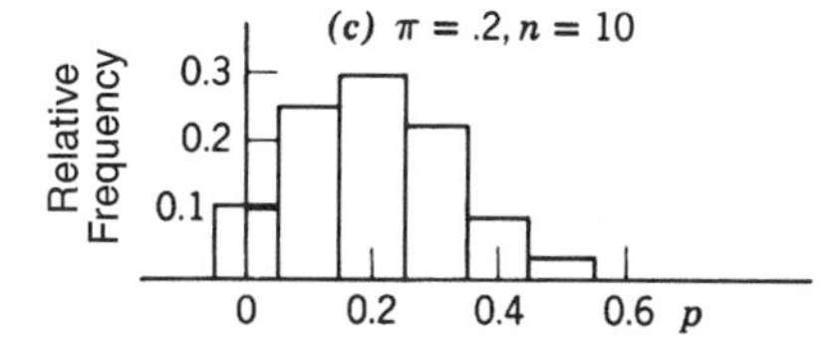

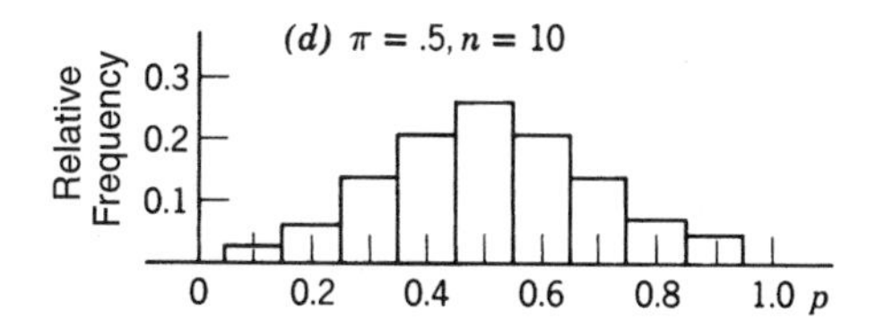

Fig. 8.3. Frequency distributions of sample proportions.

The *first* statement is that the mean of all possible sample proportions is equal to π, the population proportion. In other words, some samples give sample proportions that are higher than π; some give sample proportions that are lower than π; on the average, however, the sample proportions equal π. In symbols, $\mu_p = \pi$. In other words, p is an unbiased statistic. To verify that $\mu_p = \pi$ in Table 8.2 the mean of 100,000 p's is computed, in the same way that we compute a mean from any frequency table. It is shown to be .20; this shows that for sample size $n = 5$ and for $\pi = .2$, μ_p equals π.

The *second* statement is that the variance of the distribution of all sample proportions equals $\sigma_p^2 = \pi(1 - \pi)/n$. For continuous measurements, we have shown in Section 4.4 that the variance of the sample mean is equal to the variance of the observations divided by the sample size n. We have already shown in the cleft palate example in this chapter that the variance of an observation from a population of size 20 is $\pi(1 - \pi)$ or $\sigma^2 = \pi(1 - \pi)$. From the calculations using Table 8.2, it can be seen that the computed variance of p is equal to $\pi(1 - \pi)/n$ or $.2(1 - .2)/5 = .032$.

8.3 THE NORMAL APPROXIMATION TO THE BINOMIAL

When the sample size is large, one more statement can be made about the distribution of sample proportions: It is well approximated by a normal distribution with the proper mean and standard deviation. In this text, no use is made of binomial tables; instead the normal tables are always used in problems involving proportions.

8.3.1 Use of the Normal Approximation to the Binomial

The normal tables are used as an approximation to the binomial distribution when n, the sample size, is sufficiently large, and the question naturally arises as to what "sufficiently large" means. One rule of thumb often given is that we may use the normal curve in problems involving sample proportions whenever $n\pi$ and $n(1 - \pi)$ are both as large as or larger than 5. This rule reflects the fact that, when π is $\frac{1}{2}$, the distribution looks more like a normal curve than when π is close to 0 or close to 1; see Figure 8.3, where (*b*) looks more nearly normal than (*a*), and (*d*) looks more nearly normal than (*c*).

Use of the normal distribution approximation for the distribution of sample proportions may be illustrated by the following problem: A sample of size 25 is to be drawn from a population of adults who have been given a certain drug. In the population, the proportion of those who show a particular side effect is .2. What percentage of all sample proportions for samples of size 25 lies between .08 and .32, inclusive?

The normal approximation may be used in solving this problem since $n\pi = 25(.2) = 5$ and $n(1 - \pi) = 25(.8) = 20$ are both as large as 5. The mean of all possible p's is .2, and an approximate answer to the problem is the shaded area in Figure 8.4. For the normal curve approximation, we use the normal curve that has a mean of .2, and a standard deviation of p of

$$\sqrt{\pi(1 - \pi)/n} = \sqrt{.2(.8)/25} = \sqrt{.0064} = .08$$

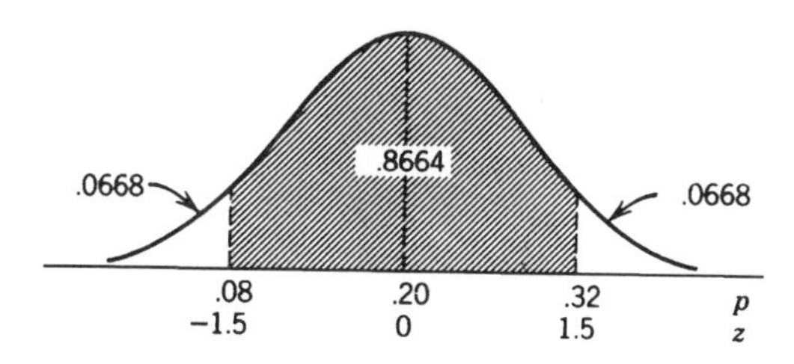

Fig. 8.4. Proportion of p from 0.08 to 0.32.

To find the area from .08 to .32 we find as usual the z corresponding to .32. Here $z = (.32 - .20)/.08 = .12/.08 = 1.5$. Using Table A.2, we find that .9332 or 93.32% of the sample proportions lie below .32, so that 6.68% of the sample proportions lie $> .32$. By symmetry, 6.68% of the sample proportions lie $< .08$. This means that about 86.64% of the sample proportions lie between .08 and .32. Thus 86.64% of sample proportions from samples of size 25 will be from .08 to .32 inclusive. The exact answer using tables of the binomial distribution is .9258; in this borderline situation with $n\pi = 5$, the normal approximation is not very close.

8.3.2 Continuity Correction

The normal approximation may be improved somewhat by using what is called a *continuity correction*. In the transformation to z, for the upper tail, a continuity correction of $1/2n$ is added to p. Since $n = 25$, the amount of the correction factor is $1/50$ or .02. For the lower tail, the same continuity correction is subtracted from p. In the example of the previous paragraphs,

$$z = \frac{(.32 + .02) - .2}{.08} = \frac{.14}{.08} = 1.75$$

From Table A.2, about .9599 of sample proportions lie $< .34$. In this example, this is equivalent to saying that .9599 of the sample proportions are $\leq .32$, inasmuch as it is impossible with a sample of size 25 to obtain a p between .32 and .36. Then, $1 - .9599 = .0401$ or 4.01% of the sample proportions are $> .32$ and by symmetry 4.01% of the sample proportions are $< .08$. So 91.98% of the sample proportions lie from .08 to .32 inclusive. Note that using the continuity correction factor resulted in an area closer to .9258, which was the exact answer from a table of the binomial distribution.

In general, the formula can be written as

$$z = \frac{(p - \pi) \pm 1/2n}{\sqrt{\pi(1 - \pi)/n}}$$

If $(p - \pi)$ is positive, then the plus sign is used. If $(p - \pi)$ is negative, then the minus sign is used. As n increases the size of the continuity correction decreases and it is seldom used for large n. For example, for $n = 100$, the correction factor is $1/200$ or 0.005 and even for $n = 50$, it is only 0.01.

8.4 CONFIDENCE INTERVALS FOR A SINGLE POPULATION PROPORTION

Suppose we wish to estimate the proportion of patients given a certain treatment who recover. If in a group of 50 patients assigned the treatment, 35 out of the 50 recover, the best estimate for the population proportion who recover is $35/50 = .70$, the sample proportion.

We then compute a confidence interval in an attempt to show where the population proportion may actually lie.

The confidence interval is constructed in much the same way as it was constructed for μ, the population mean, in Section 6.1.2. The sample proportion p is approximately normally distributed, with mean π and with standard deviation $\sigma_p = \sqrt{\pi(1-\pi)/n}$, and in order to make a 95% confidence interval just as for the population mean, we would use $p \pm 1.96\sigma_p$. Unfortunately, the standard deviation of p is unknown, so we must estimate it in some way. The best estimate that we have for π is $p = .70$, so as an estimate for σ_p, we use

$$\sqrt{p(1-p)/n}$$

or

$$\sqrt{.70(1-.70)/50} = \sqrt{.21/50} = \sqrt{.0042} = .065$$

The 95% confidence interval is then $.70 \pm (1.96)(.065)$, or $.70 \pm .13$, or .57 to .83. We are 95% confident that the true recovery rate is somewhere between 57 and 83%. The formula for the 95% confidence interval is

$$p \pm 1.96\sqrt{p(1-p)/n}$$

If we desire to include the continuity correction of $1/2n = .01$, then we would add .01 to .13 and the 95% confidence interval is $.70 \pm .14$, or .56 to .84. The formula for the 95% confidence interval with the correction factor included is

$$p \pm \left(1.96\sqrt{p(1-p)/n} + \frac{1}{2n}\right)$$

At this point, a question is often raised concerning the use of p in the formula for the standard deviation of p. It is true that we do not know π and therefore do not know the exact value of the standard deviation of p. Two remarks may be made in defending the confidence interval just constructed, in which $\sqrt{p(1-p)/n}$ has been substituted for $\sqrt{\pi(1-\pi)/n}$. First, it has been established mathematically that if n

is large and if confidence intervals are constructed in this way repeatedly, then 95% of them actually cover the true value of π. Second, even though the p that is used for π in estimating the standard deviation of π happens to be rather far from the true value of π, the standard deviation computed differs little from the true standard deviation. For example, if $\pi = .5$,

$$\sigma_p = \sqrt{.5(.5)/50} = \sqrt{.25/50} = \sqrt{.0050} = .071$$

This is not very different from the approximation of .065 that we obtained using $p = .70$. Note also that σ_p is larger for $p = .50$ than it is for any other value of p and σ_p gets progressively smaller as p gets closer to 0 or 1.

8.5 CONFIDENCE INTERVALS FOR THE DIFFERENCE IN TWO PROPORTIONS

Often two populations are being studied, each with its own population proportion, and we wish to estimate the difference between the two population proportions.

For example, it may be of interest to compare the recovery rate of patients with a certain illness who are treated surgically with the recovery rate of patients with this illness who are treated medically. Or the comparison of the proportion of women who have a certain illness with the proportion of men with the same illness may be made. Or the comparison of the proportion of diabetic amputees who succeed in walking with prostheses with the proportion of nondiabetic amputees who succeed may be of interest.

Suppose we wish to perform a clinical trial comparing two cold remedies. The first population consists of all patients who might have colds and are given the first treatment; let π_1 be the proportion of the first population who recover within 10 days. Similarly, the second population consists of all patients who might have colds and are given the second treatment; π_2 is the proportion of the second population who recover within 10 days. We take 200 patients, divide them at random into two equal groups and give 100 patients the first treatment and 100 patients the second treatment. We then calculate the proportion in each sample who recover within 10 days, p_1 and p_2, and compute the difference $p_1 - p_2$.

It is now necessary to consider the distribution of $p_1 - p_2$. In Section 6.5 on the difference between two means for continuous data, we noticed that if $\overline{X}_1$ is normally distributed with mean $= \mu_1$ and with variance $= \sigma^2_{\overline{X}_1}$, and if $\overline{X}_2$ is normally distributed with mean $= \mu_2$ and with variance $= \sigma^2_{\overline{X}_2}$, then $\overline{X}_1 - \overline{X}_2$ is normally distributed with mean $= \mu_1 - \mu_2$ and variance $= \sigma^2_{\overline{X}_1} + \sigma^2_{\overline{X}_2}$.

The situation is almost the same with the difference of two sample proportions when the sample sizes n_1 and n_2 are large.

If p_1 is normally distributed with mean $= \pi_1$, and variance $= \pi_1(1 - \pi_1)/n_1$, and if p_2 is normally distributed with mean $= \pi_2$ and variance $= \pi_2(1 - \pi_2)/n_2$, then

$p_1 - p_2$ is normally distributed with mean $= \pi_1 - \pi_2$ and variance $= \pi_1(1-\pi_1)/n_1 + \pi_2(1-\pi_2)/n_2$.

That is, the difference between two sample proportions, for large sample sizes, is normally distributed with the mean equal to the difference between the two population proportions and with variance equal to the sum of the variances of p_1 and p_2.

In our example, if 90 patients of the 100 patients who receive the first treatment recover within 10 days, $p_1 = .90$; if 80 of the 100 patients who receive the second treatment recover within 10 days, then $p_2 = .80$. Then, $p_1 - p_2 = .10$ is the best estimate for $\pi_1 - \pi_2$. Since the standard deviation of $p_1 - p_1$ is equal to

$$\sqrt{\pi_1(1-\pi_1)/n_1 + \pi_2(1-\pi_2)/n_2}$$

the usual way of forming 95% confidence intervals gives, for $\pi_1 - \pi_2$,

$$(p_1 - p_2) \pm 1.96\sqrt{\pi_1(1-\pi_1)/n_1 + \pi_2(1-\pi_2)/n_2}$$

Again the standard deviation must be estimated from the sample, as follows, using .90 in place of π_1 and .80 in place of π_2; $\sigma_{p_1-p_2}$ is estimated by

$$\begin{aligned}\sqrt{.90(1-.90)/100 + .80(1-.80)/100} &= \sqrt{(.9)(.1)/100 + (.8)(.2)/100} \\ &= \sqrt{.0025} = .05\end{aligned}$$

The 95% confidence interval is then

$$(.90 - .80) \pm 1.96(.05) = .10 \pm .098$$

or .002 to .198.

Since the 95% confidence interval for difference in recovery rate is between 0.2% and 19.8%, it may be practically 0, or it may be nearly 20%. Because both confidence limits are positive, we decide that the difference is positive, so that the population 10-day recovery rate is higher for the first treatment than for the second. We say that the difference between the two recovery rates is statistically significant, or that it is a significant difference. In calling a sample difference such as $.90 - .80 = .10$ a significant difference, we mean that it is large enough to enable us to conclude that a population difference exists. Note that we are *not* asserting that the population difference is large or important; we merely believe that it exists. In symbols, the 95% confidence interval is

$$(p_1 - p_2) \pm 1.96\sqrt{p_1(1-p_1)/n_1 + p_2(1-p_2)/n_2}.$$

The formula for the 95% confidence interval if a continuity correction factor is included is

$$(p_1 - p_2) \pm (1.96\sqrt{p_1(1-p_1)/n_1 + p_2(1-p_2)/n_2} + 1/2(1/n_1 + 1/n_2)).$$

8.6 TESTS OF HYPOTHESIS FOR POPULATION PROPORTIONS

If the sample size is large enough to use the normal curve, hypotheses concerning population proportions may be tested in much the same way as were hypotheses concerning population means. We will first give the tests for a single population proportion, and then the test for the difference in two population proportions.

8.6.1 Tests of Hypothesis for a Single Population Proportion

Suppose, for example, based on considerable past experience that it is known that the proportion of the people with a given infectious disease who recover from it within a specified time is .80 when treated with the standard medication. There is concern that the medication may not always have the same success rate due to changes in the virus causing the disease. We want to test for possible changes in the patients that have been treated recently and to find out whether or not $\pi = .80$. The null hypothesis is that H_0: $\pi_0 = .80$. We have data for 50 consecutive patients who have been treated within the last year for the disease, and find that 35 recovered, so that the sample proportion is $p = 35/50 = .70$. Under H_0, p is approximately normally distributed with mean .80 and with standard deviation

$$\sigma_p = \sqrt{\pi_0(1-\pi_0)/n} = \sqrt{.8(.2)/50} = \sqrt{.16/50} = .0566$$

Since we wish to know whether π equals or does not equal .80, we will use a two-sided test.

We wish to calculate P, the probability that a sample proportion p will be .70 or less or will be .90 or more if actually $\pi = .80$. We need the shaded area under the normal curve in Figure 8.5. First, we find the z corresponding to $p = .70$ using the following formula:

$$z = \frac{p - \pi_0}{\sigma_p}$$

or

$$z = \frac{.70 - .80}{.0566} = \frac{-.10}{.0566} = -1.77$$

From Table A.2, we find that the area below $z = +1.77$ is .9616, so that the area of the two shaded portions is $2(1 - .9616) = .0768$. That is, $P = .0768$, and we accept the null hypothesis using a significance level of $\alpha = .05$. In other words, we decide that the recovery rate for this disease may not have changed.

We may prefer a one-sided test. Possibly we may really believe that the recovery rate is lower now than in the past; then we are interested in finding an answer to the question: Is the population proportion $< .80$? We will test the null hypothesis that H_0: $\pi_0 \geq .80$. We then compute the probability in the lower side of the distribution only, so that we have $P = .0384$, and we reject the null hypothesis that $\pi \geq .80$ at

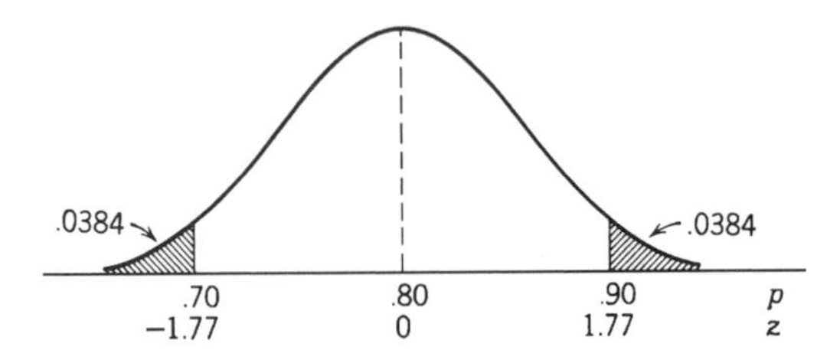

Fig. 8.5. Two-sided test of H_0: $\pi = .80$.

the .05 level. Our conclusion is that the new recovery rate is lower than the recovery rate for the past.

It should be noted that from the same set of data, two different tests have led to different conclusions. This happens occasionally. Clearly, the test to be used should be planned in advance; one should not subject the data to a variety of tests, and then pick the most agreeable conclusion. In the example, we picked the test according to the question we wished to answer.

8.6.2 Testing the Equality of Two Population Proportions

When two sample proportions have been calculated, one from each of two populations, then often, instead of giving a confidence interval for the difference in population proportions, a test is made to determine if the difference is 0.

The example of the doctor with 100 patients under each of two treatments will be used to illustrate the test procedure. In this example, the two sample proportions were $p_1 = .90$ and $p_2 = .80$, and the sample sizes were $n_1 = n_2 = 100$.

The null hypothesis is H_0: $\pi_1 = \pi_2$, and, we are merely asking whether or not the two recovery rates are the same, so a two-sided test is appropriate. We choose $\alpha = .05$.

Using the fact that $p_1 - p_2$ has a normal distribution with mean 0 ($\pi_1 - \pi_2 = 0$) if H_0 is true, we wish to find P, the shaded area in Figure 8.6, which represents the chance of obtaining a difference as far away from 0 as is .10 if actually the true mean difference is 0.

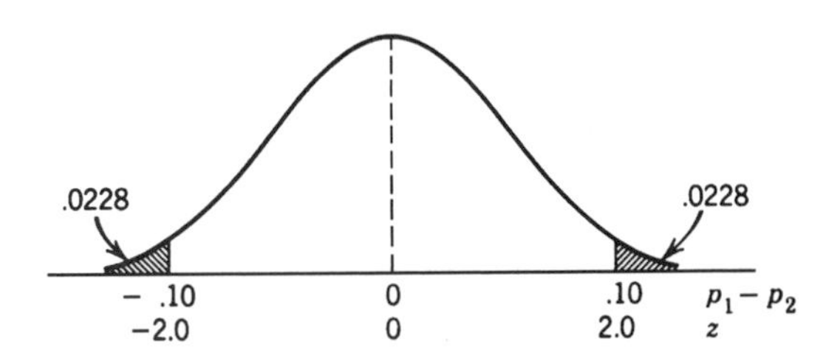

Fig. 8.6. Two-sided test of H_0: $\pi_1 = \pi_2$.

To change to the z-scale in order to use the normal tables, we need to know the standard deviation of $p_1 - p_2$. This is

$$\sqrt{\pi_1(1-\pi_1)/n_1 + \pi_2(1-\pi_2)/n_2}$$

which is unknown since the population proportions are unknown. It must be estimated.

Under the null hypothesis, π_1 and π_2 are equal, and the best estimate for each of them obtainable from the two samples is $170/200 = .85$ since in both samples combined, $90 + 80 = 170$ of the 200 patients recovered. This combined estimate can be called $\overline{p}$. It is a weighted average of the p_1 and p_2. That is

$$\overline{p} = \frac{n_1 p_1 + n_2 p_2}{n_1 + n_2}$$

When $n_1 = n_2$, $\overline{p}$ is simply the average of p_1 and p_2.

The estimate of the standard deviation is

$$\sigma_{p_1-p_2} = \sqrt{\overline{p}(1-\overline{p})/n_1 + \overline{p}(1-\overline{p})/n_2}$$

or

$$\sigma_{p_1-p_2} = \sqrt{\overline{p}(1-\overline{p})(1/n_1 + 1/n_2)}$$

In our example, $\sigma_{p_1-p_2}$ is estimated by

$$\sigma_{p_1-p_2} = \sqrt{.85(1-.85)(1/100 + 1/100)} = \sqrt{2(.1275)/100} = .050$$

Now, since we have an estimate of the $\sigma_{p_1-p_2}$, we can calculate z:

$$z = \frac{(p_1 - p_2) - (\pi_1 - \pi_2)}{\sigma_{p_1-p_2}} = \frac{(p_1 - p_2)}{\sqrt{\overline{p}(1-\overline{p})(1/n_1 + 1/n_2)}}$$

In the example,

$$\frac{.10 - .00}{.05} = 2$$

From Table A.2, the area in the upper tail is $1 - .9772 = .0228$. For a two-sided test, we have

$$P = 2(.0228) = .0456$$

With $\alpha = .05$, the null hypothesis is rejected and it is decided that the two treatments differ in their recovery rates.

If we wish to include a correction for continuity, then $1/2(1/n_1+1/n_2)$ is subtracted from a positive value of $(p_1 - p_2)$ or added to a negative value of $(p_1 - p_2)$. Here, $(p_1 - p_2) = +.10$ and $1/2(1/100 + 1/100)$ or 0.01 would be subtracted from .10.

Although usually a two-sided hypothesis is used, it is also possible to test a one-sided hypothesis. The procedure is analogous to that given in Section 7.2.2, where tests for the differences in two means were discussed. If the null hypothesis $H_0{:}\pi_1 \leq \pi_2$ is tested, then the entire rejection region is in the upper tail of the z distribution.

8.7 SAMPLE SIZE FOR TESTING TWO PROPORTIONS

The test of whether two population proportions are equal is one of the more commonly used tests in biomedical applications. Hence, in planning a study finding the approximate sample size needed for this test is often included in proposals. In order to determine the sample size, we will need to decide what levels of α and β to use so that our chances of making an error are reasonably small. Often, α is chosen to be .05 and β is chosen to be .20, but other values should be taken if these values do not reflect the seriousness of making a Type I or Type II error. For $\alpha = .05$, $z[1 - \alpha/2] = 1.96$ is used for a two-sided test (see Section 7.5). For $\beta = .20$, $z[1 - \beta] = .842$. We will also have to estimate a numerical value for π_1 and for π_2. Suppose we plan to compare performing an operation using endoscopy versus the conventional method. The proportion of complications using the conventional method is known from past experience to be approximately $\pi_1 = .12$. We wish to know whether the use of endoscopy changes the complication rate (either raises it or lowers it), and so we plan on performing a two-sided test. We certainly want to detect of difference if $\pi_2 = .20$ since that would imply that the new method is worse than the conventional method. Also, by taking a value of π_2 that is closer to one-half, we will obtain a conservative estimate of the sample size since values of π_2 closer to one-half result in a larger variance. The next step is to calculate

$$\overline{\pi} = \frac{\pi_1 + \pi_2}{2} = \frac{.12 + .20}{2} = .16$$

An approximate estimate of the needed sample size, n, in each group is computed from the following formula:

$$n = \frac{\left[z[1-\alpha/2]\sqrt{2\overline{\pi}(1-\overline{\pi})} + z[1-\beta]\sqrt{\pi_1(1-\pi_1) + \pi_2(1-\pi_2)}\right]^2}{(\pi_1 - \pi_2)^2}$$

Substituting in the numbers for the example,

$$n = \frac{\left[1.96\sqrt{2(.16)(.84)} + .842\sqrt{.12(.88) + .20(.80)}\right]^2}{(.12 - .20)^2}$$

or

$$n = \frac{[1.96\sqrt{.2688} + .842\sqrt{.2656}]^2}{.0064} = \frac{[1.01618 + .433936]^2}{.0064} = 328.6$$

which is rounded up to $n = 329$ observations in each treatment group. For small n, an approximate continuity correction factor is often added so that the corrected n' is

$$n' = n + \frac{2}{|\pi_1 - \pi_2|}$$

where $|\pi_1 - \pi_2|$ denotes the positive difference between π_1 and π_2. In our example,

$$n' = 329 + \frac{2}{.08} = 329 + 25 = 351$$

One source of confusion in estimating sample size for differences in proportions during the planning stage of the study is that there are several slightly different formulas in use as well as different correction factors. In the main, they lead to similar estimates for n but there is not precise agreement between various tables and statistical programs.

For a one-sided test with $\alpha = .05$, we use $z[1 - \alpha]$ or 1.645.

8.8 DATA ENTRY AND ANALYSIS USING STATISTICAL PROGRAMS

When entering categorical data into the computer where there are only two outcomes, subsequent data screening is simplified if a consistent numbering system is used for coding the data. For example, if we asked a series of questions where the answer was always yes or no, we could code a yes = 1 and a no = 2 for each question. If the categorical data had more than two responses, for example, a rating system for seriousness of illness, with the possible responses being mild, moderate, or severe, then we could code them mild = 1, moderate = 2, and severe = 3. The results for patients who said they did not know how severe their illness is could be coded do not know = 8 and a missing value could be coded as a blank or some other value depending on the statistical program used. By using a consistent code, errors in data entry such as someone typing a value of 4 through 7 can be easily found.

Any question with numerous missing values or do not know responses can be examined to see if there was something wrong with the question.

Note that in many computer programs a variable that has numerous missing values can result in the data for many patients being excluded even through that particular variable is not used in the current analysis. To avoid this loss of patients in the analysis, the variable with the missing values should be explicitly noted as not used or dropped.

In general, most computer programs do not calculate the confidence limits or tests of hypothesis for a single proportion described in this chapter. They do, however, provide for a count of the number of observations that occur in each category of categorical variables, and then compute proportions or percents. They will also make the graphical displays described at the start of this chapter.

If proportions for two groups are desired, then a variable that differentiates between the two groups is needed. For example, a variable could be created that is assigned the value 1 for one group and 2 for the other group. Counts can then be done by group. A test will be given in Chapter 9 that tests whether two proportions are equal. This test is widely available in computer programs.

PROBLEMS

8.1 Table A.1 lists random digits between 0 and 9. Assume that an even number is a success and an odd number is a failure. Draw three samples of size 10 and write down the proportion of successes. Make a frequency table with all the sample proportions from the entire class. Compute the mean and variance of all the sample proportions. What do you expect the population π to equal? Using π what is the variance of an observation? What is the variance of the mean of 10 observations? How does this compare with the result from the frequency table?

8.2 In a statistical report, the statement is made that the 95% confidence interval for the percentage of babies who are boys is 50.2 to 50.8%. Which of the following interpretations of this statement are true and which are false?

- **(a)** That 95% of the sample will have between 50.2 and 50.8% boys.
- **(b)** That 95% of future random samples of babies will have between 50.2 and 50.8% boys.
- **(c)** If confidence intervals are computed in the same way for each of a large number of samples of the same size, 95% of such intervals will overlap the interval 50.2 to 50.8%.
- **(d)** The mean percentage of boys will be between 50.2 and 50.8, 95% of the time.
- **(e)** That 95% of the time we would obtain $50.5 \pm .3$. If a large number of samples is taken, the interval will capture about 95% of the sample proportions.
- **(g)** If a large number of samples of the same size is taken and confidence intervals computed in the same way for each, 95% of such intervals will cover the population percentage of boys.

8.3 If the proportion of people with a certain illness who recover under a given treatment is actually .8, what percentage of sample proportions obtained from repeated samples of size 64 would be between .7 and .9?

8.4 The proportion of patients who have an allergic reaction to a certain drug is unknown. What percentage of samples of size 25 would give sample proportions that differ from the population proportion by $< .15$? Use $\pi = .5$ for estimating the standard deviation.

8.5 If the proportion of patients who recover from a certain illness under a given treatment is actually .9, what is the probability that $< 75\%$ of a sample of 100 patients recover?

8.6 One hundred patients are available to participate in an experiment to compare two cold remedies. As in the example in the text, the physician divides them into two equal groups, gives each treatment to 50 patients, and finds that the percentages of patients who recover within 10 days are 90 and 80%.

(a) Give a 95% confidence interval for the difference between the population recovery rates. What can you conclude from this confidence interval?
(b) Compare the interval in (a) with the one in the text and comment.

8.7 An experiment was performed to determine if a new educational program given by pharmacists could reduce the prescribing of a widely used analgesic that had been proven to be inferior to aspirin. The physicians were randomly assigned to either control ($n = 142$) or the new educational program ($n = 140$), and their prescriptions of the analgesic were monitored from medical records. After the education program, 58 of the physicians in the control group prescribed the analgesic and 42 in the education group. Test for no difference in the proportion prescribing the analgesic at the $\alpha = .05$ level.

8.8 In high schools A and B, the proportion of teenagers smoking has been estimated to be .20. The same antismoking educational program will be used again in high school A but a new program is planned for high school B. There is concern whether the new program will be better or worse than the old one or if the results will be about the same. A study is planned to compare the proportion smoking in the two high schools in June after the programs have been run. The investigators want to take an adequate sample of students and ask them questions concerning their smoking. They want to make sure their sample size is large enough so they can detect a difference of .10 in the proportion smoking with $\alpha = .05$ and $\beta = .10$. What size sample should they take in each high school?

8.9 The proportion of 200 children age 19–35 months from low poverty homes who had inadequate polio vaccinations was found to be 0.21 in community A. Community B, which has a special program in place to increase vaccination of children from families in poverty, measured their vaccination proportion and found it to be .12 among 150 children. Test the null hypothesis of equal proportions in the two communities using $\alpha = .05$. Also compute a 95% confidence limit for the difference in the two proportions. Contrast what statements can be made from the test and the confidence limits.

REFERENCES

Cleveland, W. S. [1985]. *The Elements of Graphing Data,* Monterey, CA: Wadsworth, 264.

Dixon, W. J. and Massey, F. J. [1983]. *Introduction to Statistical Analysis,* New York: McGraw-Hill Book Company, 281–284.

Fisher, L. D. and van Belle, G. [1993]. *Biostatistics: A Methodology for the Health Sciences,* New York: Wiley-Interscience, 178–185.

Mould, R. F. [1995]. *Introductory Medical Statistics,* 3rd Ed., Philadelphia: Institute of Physics Publishing, 67–74.

CHAPTER NINE

Categorical Data: Analysis of Two-Way Frequency Tables

In Chapter 8, we discussed the analysis of categorical data using proportions. In this chapter, the analysis will be presented for categorical data that have been summarized in the form of a two-way frequency table. By a two-way table we mean a table in which counts are displayed in rows and columns. The analyses presented here are widely used in biomedical studies. They are relatively simple to do and are available both in spreadsheet and statistical programs.

This chapter is divided into five main sections. Section 9.1 describes the different types of two-way tables and how different study designs lead to different types of tables. Section 9.2 describes two commonly used descriptive measures that can be calculated from frequency tables: relative risk and odds ratio. Section 9.3 describes the use of the chi-square test for frequency tables with two rows and two columns. Chi-square analysis of larger tables (more row or columns or both) is given in Section 9.4. Section 9.5 concludes with some brief remarks.

9.1 DIFFERENT TYPES OF TABLES

Since the analyses performed and the inferences drawn depend on what type of samples were taken, we will first discuss tables in terms of the type of study and the sampling used to obtain the counts.

9.1.1 Tables Based on a Single Sample

A common type of table is based on data from a *single sample* concerning two categorical variables. For example, in a hypothetical survey of respondents in their fifties, it was determined which respondents were current smokers (yes, no) and which have or did not have low vital capacity (a measure of lung function). Here, smoking could be considered the exposure or risk variable and low vital capacity the outcome or disease variable. The total number of respondents is $n = 120$; the counts are given in Table 9.1A.

In tables in which observations are from a single sample, the purpose of the analysis is to determine whether the distribution of results on one variable is the same regardless of the results on the other variable. For example, in Table 9.1A, is the outcome of low vital capacity in a person in their 50's the same regardless of whether the person smokes or not? From a statistical standpoint, the question being asked is whether low vital capacity is independent of smoking or is there an association between smoking and low vital capacity in this age group.

In Table 9.1A, rows provide information on vital capacity outcomes and columns provide information on smoking status. The sums over the two rows are added and placed in the total row showing 30 smokers and 90 non-smokers. Likewise, the sums over the two columns show the number with and without low vital capacity (21 and 99). The overall sample size can be computed either from the sum of the row totals or the sum of the column totals. The locations in the interior of the table that include the frequencies 11, 10, 19, and 80 are called *cells*.

In Table 9.1B, symbols have been used to replace the counts given in Table 9.1A. The total sample size is n; $a+b$ responders have low vital capacity and $a+c$ responders smoke. Only a responders have low vital capacity and also smoke.

In Table 9.1A there is only one sample of size 120. All the frequencies in the table are divided by $n = 120$ to obtain the proportions given in Table 9.2. We can see that .25 or 25% of the patients smoked and .175 or 17.5% had low vital capacity. About two-thirds of the patients neither smoked nor had low vital capacity. No matter whether counts or percentages are displayed in the table, it is difficult to see whether there is any association between smoking and low vital capacity. Descriptive statistics and tests of hypotheses are thus needed.

Graphs such as the pie charts and bar charts introduced in Section 8.1.1 are often useful in displaying the frequencies. Separate pie charts for smokers and for non-

Table 9.1. Illustrating Counts (A) and Symbols (B): Single Sample

A. Association between Smoking and Low Vital Capacity: Counts

	Smoking		
Low Vital Capacity	Yes	No	Total
Yes	11	10	21
No	19	80	99
Total	30	90	120

B. Association between Smoking and Vital Capacity: Symbols

	Smoking		
Low Vital Capacity	Yes	No	Total
Yes	a	b	$a+b$
No	c	d	$c+d$
Total	$a+c$	$b+d$	n

Table 9.2. Proportions of Total Sample from Smoking and Vital Capacity Frequencies

	Smoking		
Low Vital Capacity	Yes	No	Total
Yes	.092	.083	.175
No	.158	.667	.825
Total	.250	.750	1.000

smokers would be displayed side by side and the proportion with low vital capacity and normal vital capacity would be given in each pie. Similarly, bar graphs (see Fig. 8.2) for smokers and non smokers could be placed in the same figure. The height of the bars could be either the number or the percentage of patients with low and normal vital capacity.

9.1.2 Tables Based on Two Samples

Alternatively, the data displayed in a two-way table can be from *two samples*. For example, the categorical data might be from a clinical trial where one group of patients had been randomly assigned to a new experimental treatment and the other group of patients had been assigned to the standard treatment. These two groups of patients will be considered to be two samples. The outcome of the two treatments can be classified as a success or a failure. Table 9.3 displays the outcome of such a trial where 105 patients received the experimental treatment and 100 the standard treatment.

If proportions are computed from the results in Table 9.3, they are obtained by treatment groups. From Table 9.3, we compute the proportion of patients who were treated successfully in the experimental group as $95/105 = .905$ and the proportion successfully treated in the standard group as $80/100 = .800$. The proportions that were failures are .095 in the experimental group and .200 in the standard group. If the proportions are computed by treatment groups, they are reported the same as in Section 8.6.2.

In Chapter 8, we presented statistical methods for examining differences in the population proportions. Confidence intervals for $\pi_1 - \pi_2$ were given as well as a two-sided test of hypothesis that H_0: $\pi_1 = \pi_2$ and a similar one-sided test of hypothesis. Another method of testing the two-sided hypothesis given in Section 8.6.2 will be

Table 9.3. Outcomes for Experimental and Standard Treatment: Two Samples

	Treatment		
Outcome	Experimental	Standard	Total
Success	$95 = a$	$80 = b$	$175 = a + b$
Failure	$10 = c$	$20 = d$	$30 = c + d$
Total	$105 = n_1$	$100 = n_2$	$205 = n$

Table 9.4. Exposure to Risk Factor for a Matched Case/Control Study

	Control		
Case	Yes	No	Total
Yes	$65 = a$	$18 = b$	83
No	$7 = c$	$10 = d$	17
Total	72	28	$100 = n$

presented; it can also be used when there are more than two groups and/or more than two outcomes. The tests for equal population proportions are often called tests of *homogeneity*.

9.1.3 Tables Based on Matched or Paired Samples

In some biomedical studies, each individual subject is paired with another subject. For example, in case/control studies a control subject is often paired with a case. Table 9.4 presents an example from a hypothetical matched case/control study.

In Table 9.4, there are n pairs of matched cases and controls. In 65 instances, both the case and the control were exposed to the risk factor and in 10 instances they both were not exposed. So for 75 pairs the results were a tie. For 18 pairs the case was exposed and the control was not, and for 7 pairs the control was exposed and the case was not. Thus, among the nontied pairs, the cases were exposed more than the controls. Note that the same symbols have been used to depict the counts in Table 9.1B as in Table 9.4 but what is counted is different. Here, ties and nonties for matched pairs are given.

Note that the number of ties has no effect on the *differences* between p_1, the proportion of cases exposed to risk, and p_2, the proportion of controls exposed to risk. From Table 9.4, $p_1 = (65 + 18)/100 = .83$ and $p_2 = (65 + 7)/100 = .72$, so that the difference in proportions is $.83 - .72 = .11$. As expected, ignoring ties $18/100 - 7/100 = .11$.

When two-way frequency tables are presented that simply give the frequencies without the row and column labels, one cannot possibly know how they should be interpreted. It is essential that tables be clearly labeled in order that readers can understand them.

9.1.4 Relationship between Type of Study Design and Type of Tables

In Chapter 1, we introduced the major types of studies used in the biomedical field. Studies were divided into those in which treatment was assigned (experiments) and observational studies in which exposure or risk factors were compared to outcomes such as disease. In performing an *experiment*, we want to study the effect of a treatment in a laboratory experiment, a clinical trial, or a field trial. In these experiments, the two-way table obtained from the results typically fits the *two* sample type of table.

We will assume that the sample size in the two treatment groups is fixed and we will not be interested in estimating it. In studies of surgical treatments performed on animals, sometimes the experimental surgical treatment is performed on one side of each animal and the control treatment is performed on the opposite side. Here, the outcome for the two sides of the animal are compared to each other. In this case, the sample consists of matched pairs.

In *observational* studies, the sample is often considered to be a single sample or two fixed samples. In *surveys*, we usually take a single sample and thus compare two variables measured in the survey. We assume that the row and column proportions estimate the proportions in the population being sampled. For example, we would assume that the proportion of current smokers in Table 9.1A is an estimate of the proportion in the population of respondents we sample from and that the proportion of respondents with low vital capacity is an estimate of the proportion in the population. In other observational studies, samples may be taken from two or more groups.

In *prospective* studies, all three ways of sampling mentioned previously are possible. Sometimes a *single* sample is taken and the exposure or risk variable is measured on all the subjects during the first time period and the disease outcome is measured as it occurs. In this type of prospective sample, the proportion of subjects who are exposed or not exposed are considered to be an estimate of the proportion in the population. The heart studies (e.g., the Framingham heart study) are examples of this type of prospective study. In other prospective studies, *two* samples are taken based on exposure status (a sample of exposed and a sample of nonexposed). Here, the proportion of subjects do not represent the proportion in the population since often the exposed group is oversampled. Finally, if the same measurement is taken at two time periods in a prospective study, the data can be considered a *matched* sample. The result of measurement at time 1 could be displayed in columns 1 and 2, and the results at time 2 could be displayed in row 1 and 2 (as with the operation on two sides of each animal). In essence, a table similar to Table 9.4 is obtained.

In *case/control* or retrospective studies, there are usually far fewer cases available for study than there are controls. Almost all of the available cases are studied and only a small fraction of the controls are sampled. Hence, there are two samples(one of cases and one of controls) and the sample sizes of the cases and controls are not indicative of the numbers in the population. The number of cases and controls is considered fixed and is not a variable to be analyzed.

Case/control studies are sometimes designed with individually matched cases and controls (see Schlesselman [1982]). When this is done, the results are often displayed as given in Table 9.4 with n pairs of subjects.

Two descriptive measures will be defined and their use discussed in Section 9.2. The use of the chi-square test is discussed in Section 9.3.

9.2 RELATIVE RISK AND ODDS RATIO

In this section, relative risk and odds ratios are defined and their use is discussed. Relative risk is somewhat easier to understand but the odds ratio is used more in biomedical studies. References will be given for the odds ratio for readers who desire more information on this topic.

9.2.1 Relative Risk

In Section 8.5, we examined the effect of two different treatments upon the outcome of a medical study or the relation of an exposure or risk factor on the outcome of disease by looking at the difference between two proportions. But a difference between two proportions may not always have the same meaning to an investigator. With $p_1 = .43$ and $p_2 = .40$, the difference or $p_1 - p_2 = .03$ may appear to be quite unimportant. But with $p_1 = .04$ and $p_2 = .01$, the same difference may seem more striking. The ratio of the proportions in the first instance is $.43/.40 = 1.075$, and in the second instance $.04/.01 = 4$. This ratio is called the *relative risk* or RR.

In the example shown in Table 9.3, the proportion of failures is $p_1 = c/n_1 = 10/105$ in the experimental group and $p_2 = d/n_2 = 20/100$ in the control group. The relative risk of failure for the experimental and control group is

$$\text{RR} = \frac{10/105}{20/100} = \frac{.0952}{.20} = 0.476$$

When the numerical value of the relative risk is < 1 the relative risk is said to be negative and when the relative risk is > 1 the relative risk is said to be positive. In this case, we would say that the relative risk shows that the experimental treatment may have a negative effect on the failure rate. A relative risk of one indicates that the treatment may have no effect on the outcome.

When comparing the relative risk for an observational study, we will compare the risk of disease for the *exposed* group to the risk of disease for the *nonexposed*. For example, in the vital capacity survey (see Table 9.1A and B) we would compute

$$\text{RR} = \frac{a/(a+c)}{b/(b+d)} = \frac{11/30}{10/90} = \frac{.3667}{.1111} = 3.30$$

The relative risk for smokers for low vital capacity in their 50's is positive and 3.30 times that of non-smokers. The relative risk measures how many times as likely the disease occurs in the exposed group as compared to the unexposed group.

For prospective studies using a single sample, the resulting data would appear similar to that in Table 9.1A. The difference is that the disease outcome is measured at a later time and we would likely start the study with disease-free participants. If in a prospective study, we took a sample of smokers and a sample of non-smokers, the relative risk would also be computed in the same way. The relative risk can be used for both of these types of prospective studies.

But there are difficulties with relative risk in case/control studies. The following example illustrates the problem. Table 9.5A presents hypothetical data from an unpaired case/control study. Table 9.5B presents the same results except we have doubled the size of the control group. Note that all the numbers in the control row

Table 9.5. Illustrating Doubling Size of Control Group in (A) and (B)

A. Exposure to Risk Factor for a Unmatched Case/Control Study

Outcome	Exposure: Yes	Exposure: No	Total
Case	$70 = a$	$30 = b$	100
Control	$55 = c$	$45 = d$	100
Total	125	75	$200 = n$

B. Exposure to Risk Factor when Controls Doubled

Outcome	Exposure: Yes	Exposure: No	Total
Case	$70 = a$	$30 = b$	100
Control	$110 = c$	$90 = d$	200
Total	180	120	$300 = n$

have simply been multiplied by 2. But if we compute the relative risk from Table 9.5A we get RR $= (70/125)/(30/75) = 1.40$ and RR $= (70/180)/(30/120) = 1.56$ from Table 9.5B. This unfortunate result is due to the way the sample is taken. For a further explanation of this problem, see Schlesselman [1982].

Next, we discuss the odds ratio, a measure of the strength of a relationship that can be used for all the types of studies discussed here for two-way tables with two rows and two columns.

9.2.2 Odds Ratios

In Table 9.1A, the exposure was smoking and the disease outcome was low vital capacity. From Table 9.1A, we can estimate from our sample the odds of getting diseased. The odds are estimated as

$$\text{Odds} = \frac{\text{total diseased}}{\text{total sample} - \text{total diseased}}$$

or

$$\text{Odds} = \frac{21}{120 - 21} = \frac{21}{99} = .212$$

If the odds are equal to 0.5, we say that we are equally likely to get the disease or not. Odds can vary between 0 and ∞ (i.e., infinity).

Odds are frequently quoted by sport announcers. An announcer might say that the odds are three to one that team A will defeat team B. If the teams had played together 100 times and team A had won 75 of the games, then the odds would be $75/(100 - 75) = 3$, or 3 to 1.

We can also compute the odds of low vital capacity separately for smokers and for non-smokers. The odds for smokers are $11/(30 - 11) = 11/19 = .579$. Note that $11/19$ is equivalent in symbols from Table 9.1B to a/c. For non-smokers the odds are $10/(90 - 10) = 10/80 = .125$. Here, 10/80 is equivalent to b/d in Table 9.1B.

From the odds for smokers and for non-smokers, we can compute the ratio of these odds. The resulting statistic is call the *odds ratio* (OR) or the cross-product ratio. From Table 9.1A we can divide the odds for smokers by the odds for non-smokers to obtain

$$\text{OR} = \frac{.579}{.125} = 4.63$$

In symbols from Table 9.1B,

$$\text{OR} = \frac{a/c}{b/d} = \frac{ad}{bc}$$

For the smoking and vital capacity survey results, we can say that the odds ratio of a smoker having a low vital capacity is 4.63 times that of a non-smoker. When the odds ratio is 1, we say that being exposed has no effect on (is independent of) getting the disease. If the odds ratio is < 1, the exposure results may lessen the chance of getting the disease (negative association). If the odds ratio is > 1, we say that the exposure factor may increase the chance of getting the disease (positive association). In the smoking example, the chance of having low vital capacity appears to be positively associated with smoking. The odds ratio can vary from 0 to ∞.

The odds ratio is a commonly reported statistic in biomedical reports and journals and has several desirable properties. The magnitude of the odds ratio does not change if we multiply a row or a column of a table by a constant. Hence, it can be used in case/control studies where the sample proportion is not indicative of the proportion in the population. For rare diseases, the odds ratio can be used to approximate the relative risk (see Schlesselman [1982] or Fisher and van Belle [1993]). The odds ratio does not change if rows are switched to columns and or columns to rows. It also can be computed from more complex analyses (see Afifi and Clark [1996]). For further discussion of the odds ratio, see also Rudas [1998].

One difficulty in interpreting the odds ratio is that negative relationships are measured along a scale going from 0 to 1 and positive relationships are measured along a scale going from 1 to ∞. This lack of symmetry on both sides of 1 can be removed by calculating the natural logarithm (logarithm to the base $e = 2.718\ldots$) of the odds ratio [ln(OR)]. The ln(OR) varies from $-\infty$ to $+\infty$ with 0 indicating independence.

The ln(OR) is used in computing confidence intervals for the odds ratio since the distribution of ln(OR) is closer to a normal distribution than is the distribution of the odds ratio. The population odds ratio is denoted by ω. To compute an approximate confidence interval that has a 95% chance of including the true ω from the data in Table 9.1A, we would first compute

$$\ln(\text{OR}) = \ln(4.63) = 1.5326$$

Second, we compute an estimate of the standard error (se) of ln(OR) as

$$\text{se ln(OR)} = \sqrt{\frac{1}{a} + \frac{1}{b} + \frac{1}{c} + \frac{1}{d}}$$

or numerically

$$\text{se ln(OR)} = \sqrt{\frac{1}{11} + \frac{1}{10} + \frac{1}{19} + \frac{1}{80}}$$

or

$$\text{se ln(OR)} = \sqrt{.24479} = .495$$

Next, the confidence interval for ln(OR) is given by

$$\ln(\text{OR}) \pm z[1 - \alpha/2][\text{se ln(OR)}]$$

or for a 95% confidence interval

$$1.5326 \pm 1.96(.495) = 1.5326 \pm .9702$$

or

$$.562 < \ln(\omega) < 2.503.$$

The final step is to take the antilogarithm of the end points of the confidence limit (.562 and 2.503) to get a confidence interval in terms of the original odds ratio rather than the ln(OR). This is accomplished by computing $e^{.562}$ and $e^{2.503}$. The 95% confidence interval about ω is

$$1.755 < \omega < 12.219$$

The odds for a smoker having low vital capacity are greater than those for a non-smoker, so the ratio of the odds is > 1. Further, the lower limit of the confidence limit does not include 1.

If we had a *matched* sample such as given in Table 9.4 for a case/control study, the paired odds ratio is estimated as

$$\text{OR} = b/c = 18/7 = 2.57$$

Note that the ties in Table 9.4 (a and c) are ignored. An approximate standard error of the paired odds ratio is estimated by first taking the natural logarithm of OR = 2.57

to obtain .9439 (see Schlesselman [1982]). Next, for paired samples we compute

$$\text{se}\ \ln(\text{OR}) = \sqrt{\frac{1}{b} + \frac{1}{c}} = \sqrt{\frac{1}{18} + \frac{1}{7}} = .4454$$

and 95% confidence intervals

$$\ln(\text{OR}) \pm 1.96(.4454) = .9439 \pm .8730$$

or

$$.071 < \ln(\omega) < 1.817$$

Taking the anti-logarithm of the end points of the confidence interval for $\ln(\omega)$ yields an approximate 95% confidence interval of $1.07 < \omega < 6.15$ again using $e^{.071}$ and $e^{1.817}$. Here again, the lower confidence limit is > 1 and indicates a positive association. Other methods of computing confidence intervals for matched samples are available (see Fisher and van Belle [1993]).

9.3 CHI-SQUARE TESTS FOR FREQUENCY TABLES: TWO-BY-TWO TABLES

First we present the use of chi-square tests for frequency tables with two rows and two columns. Then, the use of the chi-square tests when there are more than two rows or two columns is discussed.

9.3.1 Chi-Square Test for a Single Sample: Two-by-Two Tables

In Table 9.1A, the observed frequencies are shown from a *single* sample of 50 year olds and two measures of association (relative risk and the odds ratio) were presented that can be used in analyzing such data. We turn now to the *chi-square test for association*, a widely used test for determining whether any association exists.

The question we now ask is whether or not there is a significant association between smoking and vital capacity in 50 year olds. The null hypothesis that we will test is that smoking and vital capacity are independent (i.e., are not associated).

To perform this test, we first calculate what is called the *expected frequencies* for each of the four cells of the table. This is done as follows: If low vital capacity is independent of smoking, then the proportion of smokers with low vital capacity should equal the proportion of non-smokers. Equivalently, it should equal the proportion with low vital capacity for the *combined* smokers and non-smokers ($21/120 = .175$). In the first row of Table 9.1A, $a = 11$ and $b = 10$. We call A and B the expected frequencies for the first row, and choose them so that

$$\frac{A}{30} = \frac{21}{120} \tag{9.1}$$

and using the same reasoning,

$$\frac{B}{90} = \frac{21}{120}$$

If we multiply equation (9.1) by 30 , we have

$$A = \frac{30(21)}{120} = 5.25$$

Similarly, the value of B is

$$B = \frac{90(21)}{120} = 15.75$$

The expected values for the two other cells (C and D) can be obtained in a similar fashion

$$\frac{C}{30} = \frac{99}{120} \quad \text{or} \quad C = \frac{30(99)}{120} = 24.75$$

and

$$\frac{D}{90} = \frac{99}{120} \quad \text{or} \quad D = \frac{90(99)}{120} = 74.25$$

Note that what we do to compute the expected value is multiply the row total by the column total for the row and column each cell is in, and then divide by the total sample size. For example, for the first cell ($a = 11$) the row total is 21 for the first row and the column total is 30 for the first column and we multiplied 21 by 30 and then divided by 120, the sample total.

One should also note that the sum of the expected frequencies A and B is 21, the sum of C and D is 99, the sum of A and C is 30, and the sum of B and D is 90. That is, the sums of the expected values are the same as the sums of the observed frequencies. For a frequency table with two rows and columns, we can compute the expected value of one of the cells and we then obtain each of the three other expected values by subtraction. For example, if we compute $A = 5.25$, then we know that $B = 21 - 5.25 = 15.75$, $C = 30 - 5.25 = 24.75$, and $D = 90 - B = 90 - 15.75 = 74.25$. Thus, knowing the row and column totals and the expected value for one of the cells allows us to compute all the expected values for a table with two rows and columns. Also, knowing the row and column totals, if we know one observed frequency we can get the other observed frequencies by subtraction.

In Table 9.6, both the observed frequencies and the expected frequencies are placed in the four cells. The expected frequencies are inside the parentheses. Since the expected frequencies are what we might expect if H_0 is true, to test whether or not H_0 is true, we look at these two sets of numbers. If they seem close together, then we decide that H_0 may be true; that is, there is no significant association between

Table 9.6. Association between Smoking and Vital Capacity: Expected Frequencies

	Smoking		
Low Vital Capacity	Yes	No	Total
Yes	11(5.25)	10(15.75)	21
No	19(24.75)	80(74.25)	99
Total	30	90	120

smoking and low vital capacity. If the two sets of numbers are very different, then we decide that H_0 is false, since what was observed is so very different from what had been anticipated.

Some method is necessary for deciding whether the observed and expected frequencies are "close together" or "very different". To make this decision, the statistic called chi-square, or χ^2, is calculated. For each cell, we subtract the expected frequency from the observed frequency, square this difference, and then divide it by the expected frequency. These are summed over all the cells to obtain χ^2. We have

$$\chi^2 = \sum^{\text{all cells}} \frac{(\text{observed} - \text{expected})^2}{\text{expected}}$$

Chi-square will serve as a measure of how different the observed frequencies are from the expected frequencies; a large value of χ^2 indicates lack of agreement, a small value of χ^2 indicates close agreement between what was expected under H_0 and what actually occurred. In the example,

$$\chi^2 = \frac{(11 - 5.25)^2}{5.25} + \frac{(10 - 15.75)^2}{15.75} + \frac{(19 - 24.75)^2}{24.75} + \frac{(80 - 74.25)^2}{74.24}$$

or

$$\begin{aligned} &\frac{5.75^2}{5.25} + \frac{(-5.75)^2}{15.75} + \frac{(-5.75)^2}{24.75} + \frac{5.75^2}{74.25} \\ &= 6.298 + 2.099 + 1.336 + 0.445 = 10.178. \end{aligned}$$

The value of chi-square computed from the particular experiment is thus 10.178, and on the basis of 10.178 we must decide whether or not the null hypothesis is true.

If the experiment were repeated over and over, the chi-square calculated would vary from one time to the next. The values of chi-square have thus a sampling distribution, just as does any sample statistic. The distribution of the values of chi-square is of the general shape pictured in Figure 9.1, a skewed distribution, with of course no values of χ^2 below 0. A χ^2 value > 3.84 is expected to occur 5% of the time.

The necessary distribution has been tabled in Table A.4, and by using this table we may find P approximately. (P is the probability, if H_0 is true, of obtaining a value of chi-square at least as large as 10.178.) In Table A.4, the first column is headed d.f.

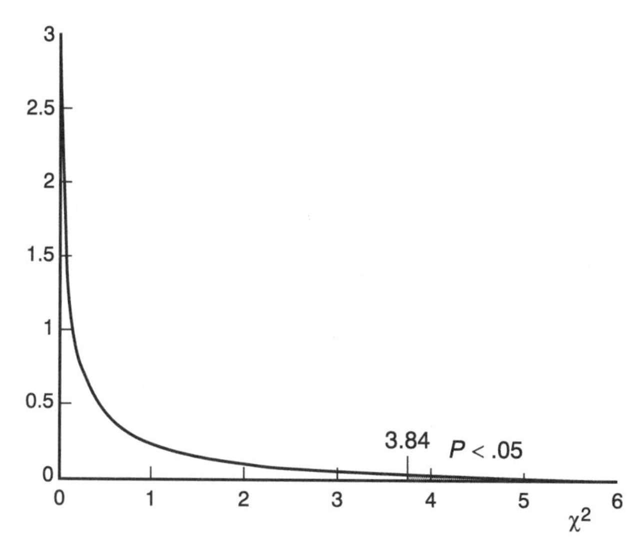

Fig. 9.1. Chi-square distribution with 1 d.f.

(degrees of freedom). The area under the χ^2 curve from 0 to $\chi^2[\lambda]$ is listed across the top of Table A.4 and the values of χ^2 are given in the body of the table. The shape of the distribution of chi-square differs for different degrees of freedom. The d.f. for the problem may be found by counting the number of independent cell frequencies in Table 9.6. That is, we count the number of cells that could be arbitrarily filled in and still keep all the same totals as in the table.

In Table 9.6, we are free to fill in frequencies in just one cell and still keep all the same totals, so that the d.f. are 1. For as soon as we know that 11 respondents smoked and had low vital capacity, then, in order to keep all the row and column totals as given in the table, we can fill in the rest of Table 9.6 by subtraction. For example, since the total number of smokers is 30, the number of smokers without low vital capacity must be $30 - 11 = 19$.

The proper d.f. to use, then, in our example is 1, and so we look for 10.178 in the first line of Table A.4. The 7.88 under .995 indicates that 99.5% of all the chi-squares are < 7.88; so our $\chi^2 = 10.178$ is expected to occur $< 0.5\%$ of the time.

If we wish to use $\alpha = .05$, we certainly would reject the null hypothesis. Even if we used an $\alpha = .01$ or .005 we would still reject the null hypothesis. We will reject the null hypothesis if the computed chi-square is greater than or equal to the tabled chi-square in Table A.4 for 1 d.f. and for our chosen value of α. Figure 9.1 depicts the value of the chi-square distribution with 1 d.f. From Figure 9.1 it is obvious that a value ≥ 10.178 is highly unlikely. That is, we will decide that there is an association between smoking and low vital capacity in 50 year olds in the population.

The chi-square test that we just made is in a sense a two-sided test. That is, we will reject the null hypothesis of smoking being independent of low vital capacity if the vital capacity of smokers was either too high or too low. The odds ratio can be used to give the direction of the association. Alternatively, examining the differences between the observed frequencies and the expected frequencies can provide information on how to interpret the results. In Table 9.6, we can see that the smokers had a higher observed number of respondents who had low vital capacity (11) than was expected (5.25), so the association is one of smoking being positively associated with low vital capacity. Examining these differences between the observed and expected is quite easy for tables with only two rows and columns but is sometimes more difficult when the table has more than two rows or columns. The analysis of tables with more than two row and/or columns will be explained in Section 9.4.

9.3.2 Chi-Square Test for Two Samples: Two-by-Two Tables

In Table 9.3, we presented data from a clinical trial where one group of patients was randomly assigned to receive the experimental treatment and the other the standard treatment. The outcome was classified as a success or a failure. Hence, the table has only two rows and two columns. There were $n_1 = 105$ patients in the experimental treatment group and $n_2 = 100$ patients in the standard treatment group. The question being asked is whether or not the treatments differ in their proportion of successes. The null hypothesis to be tested is that the proportion of successes is the same in both populations.

The total number of observed successes was 175 and the total of the two sample sizes was 205. If the two treatments both were equally effective (H_0 is true), then the expected proportion of successes in each group would be $175/205 = .8537$. Of the 105 patients given the experimental treatment, we would expect $.8537(105) = 89.63$ to be successes if the null hypothesis is true. Similarly, of the patients given the standard treatment we would expect $.8537(100) = 85.37$ to be classified as successes. The proportion of failures overall is $1 - .8537 = .1463$ (30/205 in the total column). So, in the second row we would expect $.1463(105) = 15.37$ failures in the experimental group and 14.63 failures in the standard treatment group. The results are summarized in Table 9.7.

Just as in the smoking and low vital capacity case where we had a single sample, in this clinical trial example with two samples we have to compute the expected value for only one cell and the others can be obtained by subtraction from the row and column totals. For example, for the experimental group if we compute the expected number

Table 9.7. Outcomes for Experimental and Standard Treatment: Expected Frequencies[a]

	Treatment		
Outcome	Experimental	Standard	Total
Success	95(89.63)	80(85.37)	175
Failure	10(15.37)	20(14.63)	30
Total	$105 = n_1$	$100 = n_2$	$205 = n$

[a] The numbers in parentheses are the expected values under the null hypothesis.

of successes (89.63), the expected number of failures can be obtained by subtracting 89.63 from 105. This implies that we will have 1 d.f. for testing.

In the single sample case, we computed the expected frequency by multiplying the row total by the column total, and then divided by the total sample size for any cell in the table (see Table 9.6). For example, we computed

$$A = \frac{30(21)}{120} = \frac{(a+b)(a+c)}{n}$$

For two samples, if we compute the expected frequency for A we take

$$A = \left(\frac{175}{205}\right)(105) = \frac{175(105)}{205} = \frac{(a+b)(a+c)}{n}$$

Numerically the expected frequency is computed in the same way though the null hypothesis and the method of sampling are different. From a numerical standpoint, the chi-square test for two samples is identical to that for the single sample test. (This statement holds true even when there are more than two rows and/or columns.) Again we compute

$$\chi^2 = \sum^{\text{all cells}} \frac{(\text{observed} - \text{expected})^2}{\text{expected}}$$

in the same way as before. For the frequencies in Table 9.7, we have

$$\chi^2 = \frac{(95-89.63)^2}{89.63} + \frac{(80-85.37)^2}{85.37} + \frac{(10-15.37)^2}{15.37} + \frac{(20-14.63)^2}{14.63}$$

or

$$\chi^2 = 4.50$$

With $\alpha = .05$ using Table A.4, we reject the null hypothesis of equal proportions in the two populations since the computed chi-square is larger than the tabled chi-square with 1 d.f. ($4.50 > 3.84$).

By comparing the observed and expected values in Table 9.7, we see a higher observed frequency of successes than expected in the experimental group and a lower observed frequency of successes than expected in the standard treatment group. This indicates that the experimental treatment may be better than the standard treatment.

9.3.3 Chi-Square Test for Matched Samples: Two-by-Two Tables

In Table 9.4, there are 100 pairs of individually matched cases and controls presented in a table with two rows and two columns. The appropriate null hypothesis is that there is no association between getting the disease (being a case or a control) and being exposed or not exposed to the risk factor. In other words, in the populations from which the cases and controls were sampled, the number of pairs where the case is exposed and control is not (B) is equal to the number of pairs where the control is exposed but the case is not (C). As was described in the Section 9.1.3 on tables for matched samples, the counts entered into Table 9.4 are quite different from those in the other tables, so a different χ^2 test is used.

In order to test this null hypothesis, we compute

$$\chi^2 = \frac{(b-c)^2}{(b+c)}$$

or

$$\chi^2 = \frac{(18-7)^2}{18+7} = \frac{121}{25} = 4.84$$

The computed chi-square has a chi-square distribution with 1 d.f. If we decided to use an $\alpha = .05$, then we would reject the null hypothesis since the computed value of chi-square (4.84) is greater than the tabled value (3.84) in Table A.4. In the population, we appear to have more instances where the case was exposed to the risk factor and the control was not than when the control was exposed and the case was not.

This test is commonly called McNemar's test. Note that when computing the McNemar's test, we do not use $a = 65$ or $d = 10$ in the computation. This mirrors the results in Section 9.1.3 where we showed that the differences in the proportions did not depend on the ties for paired data.

9.3.4 Assumptions for the Chi-Square Test

To use the chi-square distribution for testing hypotheses from two-way frequency table data (single sample or two sample), we need to make several assumptions. Note these assumptions apply also to any size table.

One assumption is that we have either a simple random sample from a single population or two simple random samples from two populations. Second, within each sample, the outcomes are distributed in an identical fashion. For example, if we have a sample of patients we are assuming that the chance of a successful treatment is the same for all the patients. These assumptions may not be completely met in practice.

For the matched sample chi-square test, we have to assume that a simple random sample of pairs has been taken.

Further, the sample size must be large enough to justify using the chi-square distribution. This will be discussed in Section 9.3.5 for two-by-two tables and in Section 9.4.4 for larger tables.

9.3.5 Necessary Sample Size: Two-by-Two Tables

The results given in Table A.4 for the chi-square distribution are a satisfactory approximation for testing hypotheses only when the *expected frequencies* are of sufficient size. There is some difference of opinion on the needed size of the expected values for the single sample and the two sample case. For tables with *two* rows and *two* columns, many authors say that no expected frequency should be < 5 (see Fleiss [1981]). Others say that all expected values should be > 2 or 3 (see Wickens [1989]). Small expected frequencies occur either when the overall sample size is small or when one of the rows or columns has very few observations in one of the row or column totals. For example, if a disease is rare, a prospective study where patients are followed until they get the disease may find very few diseased subjects.

Fortunately, there is a test that is widely available in statistical programs that can be used for tables with two rows and two columns when the expected frequencies are less than the recommended size. The test is called the Fisher's exact test. An explanation of this test is beyond the scope of this book (see Agresti [1996], Fleiss [1981], and particularly Wickens [1989]) on the pros and cons of using this test.

Fisher's exact test is widely used when the expected value in any cell is small. We recommend using a statistical program to perform the test since it takes considerable effort to do it otherwise. In the statistical programs, the results are often given for both a one-sided and a two-sided test. The two-sided test should be chosen if one wishes to compare the results from the exact test with those from the chi-square test. The programs will report the P values for the test. The Fisher's exact test tends to be somewhat conservative when compared to the chi-square test (see Wicken [1989]). Fisher's test should not be used with matched samples.

9.3.6 The Continuity Correction: Two-by-Two Tables

As was mentioned in Section 9.3.1, using Table A.4 results in approximate P values. For frequency tables with 1 d.f. (two rows and two columns) the approximation is poorer than for larger tables. To adjust for this approximation, a correction factor called the *Yates' correction for continuity* has been proposed. To use the correction factor, we subtract 0.5 from each of the positive differences between the observed and expected frequencies before the difference is squared. For example, if we examine Table 9.7, the differences between the observed and expected frequency for all four cells are either +5.37 or −5.37. The positive differences are all +5.37. The corrected χ_c^2 is given by

$$\chi_c^2 = \frac{(5.37 - .5)^2}{89.63} + \frac{(5.37 - .5)^2}{85.37} + \frac{(5.37 - .5)^2}{15.37} + \frac{(5.37 - .5)^2}{14.63} = 3.71$$

or

$$\chi_c^2 = \sum^{\text{all cells}} \frac{(|\text{ observed} - \text{expected}| - .5)^2}{\text{expected}}$$

The vertical line (|) is used to denote positive or absolute values. Note that for each observed minus the expected frequency, the value has been made smaller by 0.5 so the computed chi-square is smaller when the correction factor is used. In this case, the original chi-square was 4.50 and the corrected chi-square is 3.71. With large sample sizes the correction factor has less effect than with small samples. In our sample, the uncorrected chi-square had a P value of .034 and the corrected one has a P value of .054. Thus with $\alpha = .05$, we would reject the null hypothesis for the uncorrected chi-square and not reject it if we used the Yates' correction.

There is disagreement on the use of the continuity correction factor. Some authors advocate its use and others do not. The use of the continuity correction gives P values that better approximate the P values obtained from the Fisher's exact test. Using the continuity correction factor does not make the P values obtained closer to those from the tabled chi-square distribution. The result is a conservative test where the percentiles of the corrected chi-square tend to be smaller than the tabled values.

In the example just given, we suggest that both corrected and uncorrected values be given to help the user interpret the results.

For matched samples, the same correction factor can be used. Here, the formula can be simplified so that the corrected χ_c^2 is

$$\chi_c^2 = \frac{(|b - c| - 1)^2}{(b + c)}$$

where $|b - c|$ denotes the positive difference (see Fleiss [1981]).

9.4 CHI-SQUARE TESTS FOR LARGER TABLES

So far in this chapter the discussion of the use of measures of association and chi-square tests has been restricted to frequency tables with two rows and two columns. To illustrate the analysis of larger tables, we now present a hypothetical example of a two-way table with four rows and four columns. In general, frequency tables can be described as having r rows and c columns.

9.4.1 Chi-Square for Larger Tables: Single Sample

Our example is from a hypothetical health survey in which a single sample of respondents was taken from a population of adults in a county. The health survey included questions on health status, access to health care, and whether or not the respondents followed recommended preventive measures. The results from two of the questions are given in Table 9.8. The first question was, "In general, would you say your health is *excellent, good, fair, or poor*?" These four choices were coded 1, 2, 3, and 4, respectively. The second question was, "Are you able to afford the kind of medical care you should have?" The possible answers were *almost never, not often, often*, or *always* and again were coded 1, 2, 3, and 4. The data were analyzed using a statistical program. Here, the results are given in Table 9.8. Both observed frequencies and expected frequencies (in parentheses) are displayed.

The expected frequencies have been computed using the same method used for tables with two rows and columns. In each cell, the row and column totals that the

Table 9.8. Association between Health Status and Affording Medical Care

Health Status	Afford Medical Care[a] Almost Never	Not Often	Often	Always	Total
Excellent	4(8.40)	20(22.32)	21(24.72)	99(88.56)	144
Good	12(18.02)	43(47.90)	59(53.04)	195(190.04)	309
Fair	11(6.13)	21(16.27)	15(18.02)	58(64.57)	105
Poor	8(2.45)	9(6.51)	8(7.21)	17(25.83)	42
Total	35	93	103	369	600

[a] Expected frequencies are in parentheses.

cell falls in are multiplied together and the product is divided by the total frequency, or 600. For example, for the respondents who reported their health was fair and that they often could afford medical care, the expected frequency is computed as $105(103)/600 = 18.025$, which has been rounded off to 18.02 as reported in Table 9.8.

In order to test the null hypothesis that the answers to the health status question were independent of the answers to the access to health care question for adults in the county, we compute chi-square using the same formula as the earlier examples with two rows and columns; now, however, there are 16 cells rather than 4.

$$\chi^2 = \sum^{\text{all cells}} \frac{(\text{observed} - \text{expected})^2}{\text{expected}}$$

or

$$\chi^2 = \frac{(4 - 8.40)^2}{8.40} + \frac{(20 - 22.32)^2}{22.32} + \cdots + \frac{(17 - 25.83)^2}{25.83} = 30.7078$$

The d.f. are always given by $(r-1)(c-1)$ or in this example by $(4-1)(4-1) = 9$. Here r is the number of rows and c is the number of columns. The formula for the d.f. can be obtained by noting that we are free to fill in $r - 1$ times $c - 1$, or in this case 9 of the cells, and still get the same row and column totals. For example, if we know that we had 4, 20, and 21 observations in the first three columns of the first row and that the total number in the first row was 144, then we would know that the last observation in the first row must be 99.

In order to determine if the computed value of chi-square is significant, we look in Table A.4 using the row with 9 d.f. Note that the tabled value of chi-square that is needed for a small P value gets larger as the d.f. increases. For 9 d.f., we need a computed chi-square ≥ 16.92 to reject the null hypothesis of no association at the $\alpha = .05$ level; we need a tabled value of only 3.84 for 1 d.f. Since our computed value of 30.7078 is larger than the tabled value of 23.59 for the column headed .995, we know that our P value is $< .005$ or .5%.

9.4.2 Interpreting a Significant Test

We now know that there is a statistically significant association between health status and being able to afford medical care. But we are left with the problem of interpreting that association. This can be a particular problem for tables with numerous rows and columns. One way this can be done is to compare the observed and expected frequencies in Table 9.8. For example, if we look at the results for the respondents who rated their health as poor, we see that we have higher observed values than expected if there was no association for those who almost never, not often, or often could afford medical care and observed frequencies lower than expected values for those who were always able to afford the medical care they needed. At the other extreme, the respondents who rated their health as excellent had lower observed values than expected for having problems affording medical care. This type of comparison of observed and expected values can provide some insight into the results. It is easier to do when we have considerable knowledge about the variables being studied. It is often easier when the results are from treatment groups (tables with two or more treatment groups) than for survey data from a single sample. With different treatment groups, we can compare the observed and expected values by examining the treatment groups one at a time.

Another method of interpreting the results is to see what contributes most to the numerical size of the computed chi-square. Many statistical programs print out the individual terms,

$$\frac{(\text{observed} - \text{expected})^2}{\text{expected}}$$

for each cell. This option may be called cells chi-square or components of chi-square. Table 9.9 gives the cells chi-square for the data given in Table 9.8.

From Table 9.9, it can be seen that the major contribution to the overall chi-square comes from those who reported their health status as poor and almost never could afford the health care they should have. In general, the responses falling in the first column or last row make the largest contribution to the computed chi-square.

Table 9.9. Association between Health Status and Affording Medical Care: Cells Chi-Square

	Afford Medical Care				
Health Status	Almost Never	Not Often	Often	Always	Total
Excellent	2.30	0.24	0.56	1.23	4.34
Good	2.01	0.50	0.67	0.13	3.31
Fair	3.88	1.37	0.51	0.67	6.43
Poor	12.57	0.95	0.09	3.02	16.63
Total	20.77	3.07	1.82	5.05	30.71

Another method of interpreting the results in larger tables is to present bar graphs as mentioned at the end of Section 9.1.1.

9.4.3 Chi-Square Test for Larger Tables; More Than Two Samples or Outcomes

We can also analyze larger tables that have more than two samples or treatments and/or more than two outcomes. The method of computing chi-square is precisely the same as that just given for the case of a single sample. The d.f. are also the same, namely, $(r - 1)(c - 1)$. The null hypothesis is that the proportion of cases in the outcome categories in the population are the same regardless of which sample they were in (or treatment received).

If the P value is significant, the results should be interpreted. It is often useful to look at the results by the treatment group and examine what outcomes occur frequently or infrequently depending on which group they are in. Visually, bar graphs can be displayed for each group. The use of bar graphs is often the simplest way to interpret the results and this option is available in many statistical programs. Alternatively, the proportions in each group can be displayed in a two-way table.

9.4.4 Necessary Sample Size for Large Tables

The likelihood of having expected values too small in some cells to obtain an accurate estimate of the P value from Table A.4 tends to increase as the number of rows and columns in a table increases. Even if the overall sample size is large, it is possible that one or more rows or columns may contain very few individuals. For example, in biomedical applications some symptoms or test results can occur very rarely. If a row or column total is small, then often at least some of the cells in that row or column will have small expected values. Note that in computing the expected value for a cell, we multiply the row total by the column total that the cell falls in and divide by the overall n. If, for example, a row total was only 1, then the expected value would be the column total divided by n which surely is < 1.

When there are large tables (> 1 d.f.), a few cells having an expected value of about 1 can be tolerated. The total sample size should be at least four or five times the number of cells (see Wickens [1989]). The rules vary, from author to author. From a practical standpoint it does not seem sensible to have such small numbers that shifting one answer from one cell to another will change the conclusion.

When faced with too small expected values in large tables, the most common practice is to combine a row or column that has a small total with another row or column. For example, if the survey where health status and access to medical care had been performed on college students, we would probably have very few students who stated that they had poor health. If that happened, we might combine the categories of poor and fair health to get a larger row total. The choice of what to combine depends also on the purpose of the study. Rows or columns should not be combined unless the resulting category is sensible. It is sometimes better to compute an inaccurate chi-square than to disregard meaningful results. Whenever this is done, the reader of the results should be warned that the chi-square may be quite inaccurate. Usually, combinations of rows and columns can be found such that the resulting categories are worth analyzing.

9.5 REMARKS

The odds ratio and the chi-square test are frequently used in analyzing biomedical data. In particular the chi-square test is widely used in data analysis since many of the variables measured in biomedical studies are categorical or nominal data. The test is also widely available not only in statistical programs but also in some spreadsheet and other programs. The test is very easy to do but often users of the test do not put enough emphasis on explaining the results so that a reader can understand what actually happened. Simply giving a P value is usually insufficient.

PROBLEMS

9.1 In a study of weight loss, two educational methods were tested to see if they resulted in the same weight loss among the subjects. One method was called the standard method since it involved lectures on both eating and exercise. The experimental method in addition to lectures included daily reporting of dietary intake by e-mail followed by return comments on the same day. After a 1-month period, subjects were rated as to whether they met their goal for weight loss.

	Treatment		
Goal	Experimental	Standard	Total
Met	28	13	41
Not Met	24	37	61
Total	52	50	102

(a) Test whether the same proportion of the two treatment groups met their goals for weight reduction.

(b) Compute the relative risk of meeting their goal for the experimental and the control groups.

(c) Compute the odds ratio of meeting their goal for the two groups and give 95% confidence intervals.

9.2 In a case/control study, the investigators examined the medical records of the last 100 consecutive patients who had been treated for colon cancer. A control group of 100 patients from the same hospital who had been treated for abdominal hernias was used for controls. Information on current smoking status was obtained from the medical records. Of the 100 cases, 45 were current smokers and 55 were not. Of the 100 controls, 26 were current smokers and 74 were not.

(a) Compute the odds ratio of having colon cancer based on smoking status.

(b) Multiply the number of controls by 2, so there are 200 controls, 52 of whom smoke and 148 who do not smoke. Compute the same odds ratio as in (a).

(c) Compute chi-square and test that the proportion of smokers is the same among cases and controls for (a) and (b). Is the chi-square the same if the number of controls is doubled?

9.3 A matched sample case/control study was performed using cases and controls who have been matched on age and eye color and who all worked for the same chemical company. The cases had been diagnosed with melanoma. The exposure factor being studied was exposure to a chemical. The number of pairs is 75.

	Cases Exposed		
Controls exposed	Yes	No	Total
Yes	12	16	24
No	23	29	52
Total	35	40	75

(a) Compute chi-square for this matched sample case.

(b) Compute the odds ratio and its 95% confidence limit. Write in words what this odds ratio means and what this confidence limit implies.

9.4 In a health survey, adult respondents were asked if they ever had been assaulted by a domestic partner or spouse. The results were obtained for those who were currently living with their partner or spouse, and for those who were separated or divorced. There were 262 adults who reported that they were currently living with their partner or spouse and 28 of these 262 reported having been assaulted. Of the 105 adults who were currently separated or divorced, 33 reported having been assaulted. Test if there is an association between having been assaulted and currently living with your partner or spouse. Compute the odds ratio.

9.5 In a case/control study of patients with gallstones, information on coffee consumption was obtained. Among the 80 controls, 35 reported drinking at least 4 cups of coffee per day, whereas 22 of the 82 cases drank 4 or more cups. Test whether cases and controls drink the same amount of coffee.

9.6 The following table is based on samples of patients diagnosed with gastric cancer, peptic ulcer, and controls from the same health maintenance operation. From the following table, test whether there are significant differences in blood type among the three groups.

Blood Type	Gastric Cancer	Peptic Ulcer	Controls
O	400	1000	3000
A	400	700	2500
B	75	150	500
AB	25	50	200
Total	900	1900	6200

9.7 Workers were classified as having no ($n = 949$), mild ($n = 44$), moderate ($n = 13$), or severe heart failure ($n = 1$). Researchers want to compare these results with information on job history using a chi-square test. What do you suggest they do first before performing a chi-square test?

REFERENCES

Afifi, A. A. and Clark, V. [1996]. *Computer-Aided Multivariate Analysis,* 3rd Ed., London: Chapman & Hall, 285–287.

Agresti, A. [1996]. *An Introduction to Categorical Data Analysis,* New York: Wiley.

Fisher, L. D. and van Belle, G. [1993]. *Biostatistics: A Methodology for the Health Sciences,* New York: Wiley-Interscience, 197, 211.

Fleiss, J. L. [1981]. *Statistical Methods for Rates and Proportions,* 2nd ed., New York: Wiley, 25–26, 113–115.

Rudas, T. [1998]. *Odds Ratios in the Analysis of Contingency Tables,* Thousand Oaks, CA: Sage.

Schlesselman, J. J. [1982]. *Case-Control Studies,* New York: Oxford University Press, 105–123, 38, 211, 33–34.

Wickens, T. D. [1989]. *Multiway Contingency Tables Analysis for the Social Sciences,* Hillsdale, NJ: Lawrence Erlbaum Associates, 30, 44–46.

CHAPTER TEN

Variances: Estimation and Tests

In Chapter 4, we discussed the importance of variability, and by now it should be apparent that one of the major problems in working with any type of data is assessing its variability. Even if the ultimate goal is to estimate the mean of a population, we first obtain an estimate of its variance and standard deviation, and then use this estimate in testing hypotheses or developing confidence intervals for the mean.

In Chapters 6 and 7, we presented confidence intervals and tests of hypotheses for population means. This chapter gives confidence intervals for population variances and for population standard deviations as well as tests of hypotheses concerning population variances. Section 10.1 reviews how to make point estimates of a pooled variance from two samples and gives a comparable formula for the pooled estimate from more than two samples. In Section 10.2, confidence intervals for variances and standard deviations are given. Section 10.3 describes how to test whether two population variances are equal. In Section 10.4, an approximate t test is given that can be used if we find that the population variances are unequal. Finally, in Section 10.5, a brief discussion and references are given for testing the equality of two distributions when variances are unequal or the distributions are not normally distributed.

10.1 POINT ESTIMATES FOR VARIANCES AND STANDARD DEVIATIONS

First, a reminder as to how to estimate a population variance or standard deviation. From a single sample, the sample variance s^2 is calculated and used as the best point estimate of the population variance σ^2, and the sample standard deviation s is used as the point estimate for σ.

Sometimes, however, more than one sample is available for estimating a variance. In Section 6.5.3, for example, when group comparisons were made in the study of gains in weight under two diets, it was assumed that the variability of gains in weight was the same under the two diets. In other words, the population of gains in weight under the supplemented diet and the population of gains in weight under the standard diet had the same variance, σ^2. The two sample variances were pooled, and s_p^2 was used to estimate σ^2, with

$$s_p^2 = \frac{(n_1 - 1)s_1^2 + (n_2 - 1)s_2^2}{n_1 + n_2 - 2}$$

When there are several samples rather than just two for estimating a single variance, the sample variances are pooled in a similar way. In general, for k samples, the pooled estimate of the variance is

$$s_p^2 = \frac{(n_1 - 1)s_1^2 + (n_2 - 1)s_2^2 + \cdots + (n_k - 1)s_k^2}{n_1 + n_2 + \cdots + n_k - k}$$

where the variances of the k samples are $s_1^2, s_2^2, \ldots, s_k^2$ and the sample sizes are n_1, $n_2, \ldots, n_k$.

For example, if $s_1^2 = 20$, $s_2^2 = 30$, $s_3^2 = 28$, $n_1 = 12$, $n_2 = 16$, and $n_3 = 5$, then

$$s_p^2 = \frac{11(20) + 15(30) + 4(28)}{12 + 16 + 5 - 3} = \frac{782}{30} = 26.07$$

and $s_p = 5.1$.

10.2 CONFIDENCE INTERVALS FOR VARIANCES AND STANDARD DEVIATIONS

We will first present the case where a single sample is available, and then give the results for more than one sample.

When the population from which the data are drawn has a normal distribution, the chi-square table (Table A.4) may be used for confidence intervals involving a single population variance. Note that Table A.4 has already been used in Chapter 9 in the chi-square test of two-way frequency tables.

10.2.1 Confidence Intervals for the Variance and Standard Deviation: Single Sample

Consider first a single random sample of size n, drawn from a normally distributed population whose variance is σ^2. If the sample variance is s^2, then the quantity $(n - 1)s^2/\sigma^2$ could be calculated if we knew σ^2, and such a quantity would of course vary from one sample to the next and thus have a distribution. This distribution is the chi-square distribution (Table A.4) with $n - 1$ d.f. Note that the number used in the denominator in calculating s^2 equals the d.f., $n - 1$.

To illustrate the use of the chi-square table (Table A.4) in forming a confidence interval for the population variance, let us use the sample of 16 gains in weight of infants under a supplemented diet given in Table 6.1. In this example, $n = 16$, $s^2 = 20{,}392$, and $s = 142.8$ g. We decide to find 95% confidence limits and so

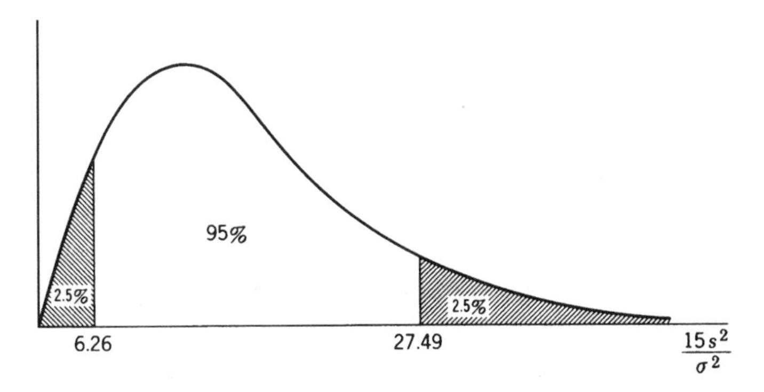

Fig. 10.1. Distribution used in confidence interval for σ^2.

$\alpha/2 = 2.5\%$ and $1 - \alpha/2 = 97.5\%$. The distribution of $15s^2/\sigma^2$ looks something like Figure 10.1, and from Table A.4 we find, looking in the row with d.f. = 15, that in .025 or 2.5% of all possible samples, $15s^2/\sigma^2$ falls below 6.26, the lower limit. In 97.5% of all samples, $15s^2/\sigma^2$ falls below 27.49, the upper limit.

In other words, in 95% of all samples, $15s^2/\sigma^2$ will lie between 6.26 and 27.49. That is,

$$6.26 < \frac{15s^2}{\sigma^2} < 27.49$$

An algebraically equivalent statement is that

$$\frac{15s^2}{27.49} < \sigma^2 < \frac{15s^2}{6.26}$$

By substituting $s^2 = 20{,}392$, we have

$$\frac{15(20{,}392)}{27.49} < \sigma^2 < \frac{15(20{,}392)}{6.26}$$

or

$$11{,}127 < \sigma^2 < 48{,}863$$

Thus 11,127 to 48,863 is a 95% confidence interval for σ^2.

Since 95% of all possible samples are such that $15s^2/\sigma^2$ lies between 6.26 and 27.49, then 95% of all the intervals formed in the manner given above will cover the population variance.

To obtain an interval for σ, the standard deviation of the population, we simply take the square root of 11,127 and of 48,863. Then, we have $105.5 < \sigma < 221.0$ g, the 95% confidence interval for σ is 105.5 – 221.0 g.

In general, the $100(1-\alpha)\%$ confidence interval for σ^2 is given by

$$\frac{(n-1)s^2}{\chi^2_{1-\alpha/2}} < \sigma^2 < \frac{(n-1)s^2}{\chi^2_{\alpha/2}}$$

Here, $\chi^2_{\alpha/2}$ denotes the lower $\alpha/2$ percentile of the chi-square distribution with $n-1$ d.f. and $\chi^2_{1-\alpha/2}$ denotes the upper $1-\alpha/2$ percentile of the chi-square distribution. The interval for σ is obtained be taking the square root as in the example.

We emphasize that the confidence intervals given above are appropriate for simple random samples from normal distributions. When the underlying distributions are skewed or include outliers, such intervals may be quite misleading. For example, if we made an error entering $X = 929$ instead of $X = 229$, the new mean would be 355.6, only 14% larger than the original mean of 311.9. But the new variance would be 43,282, over twice the size of the original variance 20,392. This change in the sample variance makes both the upper and lower confidence limits more than twice as large as they should be. Outliers such as the one just mentioned can be considered as a sample from a different population, a population of errors.

10.2.2 Confidence Intervals for Variances and Standard Deviations: More Than One Sample

When more than one sample is available to estimate a variance, then a confidence interval can be constructed using s_p^2, the pooled estimate of the variance. Again, Table A.4 can be used, since it can be shown mathematically that

$$\frac{(n_1 + n_2 + \cdots + n_k - k)s_p^2}{\sigma^2}$$

has a chi-square distribution, provided the samples are from normal populations, all with the same variance, σ^2. The d.f. to be used in entering the table are $n_1 + n_2 + \cdots + n_k - k$.

In the example given earlier in Section 10.1, with three samples and $s_1^2 = 20$, $s_2^2 = 30$, $s_3^2 = 28$, and $n_1 = 12$, $n_2 = 16$, $n_3 = 5$, the pooled estimate of the variance is calculated to be $s_p^2 = 26.07$. The d.f. are $12 + 16 + 5 - 3 = 30$, and in 95% of such experiments, $30s_p^2/\sigma^2$ lies between 16.79 and 46.98 (Table A.4), as illustrated in Figure 10.2. The 95% confidence interval for σ^2 is then

$$\frac{30s_p^2}{46.98} < \sigma^2 < \frac{30s_p^2}{16.79}$$

or substituting $s_p^2 = 26.07$ we have

$$\frac{30(26.07)}{46.98} < \sigma^2 < \frac{30(26.07)}{16.79}$$

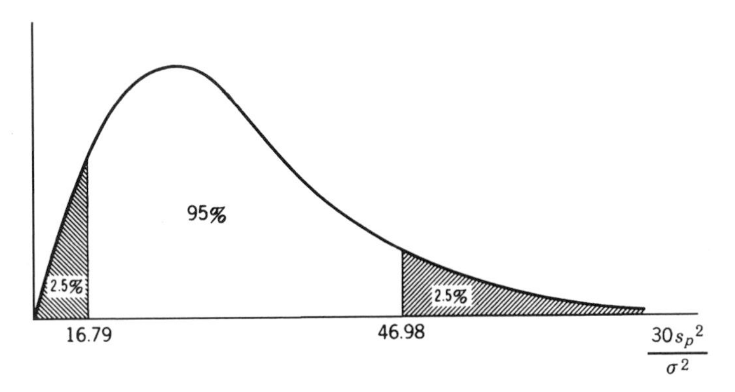

Fig. 10.2. Distribution of $30s_p^2/\sigma^2$.

giving an interval of

$$16.65 < \sigma^2 < 46.58.$$

The 95% confidence interval for σ obtained by taking square roots is 4.08–6.82.

Tests concerning population variances or standard deviations can be made by means of these confidence intervals. For the example involving three independent samples, the null hypothesis H_0: $\sigma = 2$ would be rejected at a level of $\alpha = .05$ because 2 is not included in the 95% confidence interval 4.08–6.82. Similarly, the null hypothesis H_0: $\sigma = 5$ would be accepted at $\alpha = .05$ level.

Again, we emphasize that confidence intervals using the pooled estimate of the variance are only appropriate for simple random samples from normal distributions. When the distributions are badly skewed, such intervals can be misleading.

10.3 TESTING WHETHER TWO VARIANCES ARE EQUAL: *F* TEST

Frequently, with two samples we wish to make a test to determine whether the populations from which they come have equal variances. For example, with two methods for determining blood counts, we may wish to establish that one is less variable than the other. Another situation arises when we need to pool the two sample variances and use s_p^2 as an estimate for both population variances. In making the usual t test to test the equality of two means, we assume that the two populations have equal variances, and a preliminary test of that assumption is sometimes made. This option is often found in statistical programs.

Under assumptions given below, a test of H_0: $\sigma_1^2 = \sigma_2^2$ can be made using a distribution known as the F distribution.

With two independent simple random samples of sizes n_1 and n_2, respectively, we calculate their sample variances s_1^2 and s_2^2. If the two populations from which we have sampled are both normal with variances σ_1^2 and σ_2^2, then it can be shown mathemati-

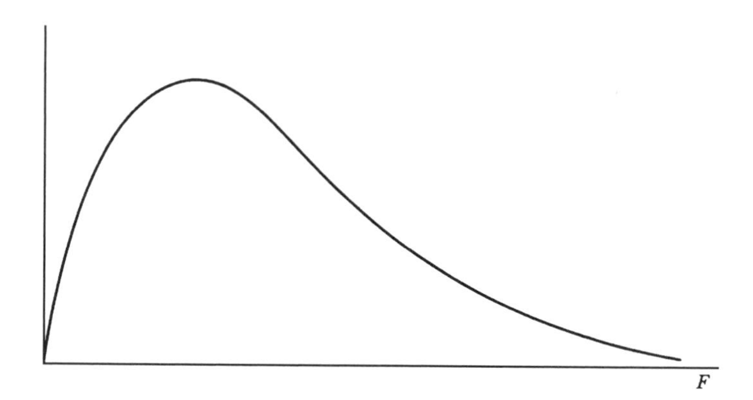

Fig. 10.3. The F distribution.

cally that the quantity $(s_1^2/\sigma_1^2)/(s_2^2/\sigma_2^2)$ follows an F distribution. The quantity in the numerator, when multiplied by its d.f., $n_1 - 1$, has a chi-square distribution; a similar statement may be made for s_2^2/σ_2^2. The exact shape of the distribution of F depends on both d.f. $n_1 - 1$ and $n_2 - 1$, but it looks something like the curve in Figure 10.3. The F distribution is available in Table A.5.

As an example of testing the equality of two variances, we return to Table 7.2 and wish to decide whether the variances are the same or different in the populations of hemoglobin levels for children with cyanotic and acyanotic heart disease. We are asking whether or not σ_1^2 is different from σ_2^2 and the null hypothesis is, H_0: $\sigma_1^2 = \sigma_2^2$. A two-sided test is appropriate here. We have $n_1 = 19$, $n_2 = 12$, $s_1^2 = 1.0167$, and $s_2^2 = 1.9898$. The test statistic we use is given by

$$F = \frac{s_1^2/\sigma_1^2}{s_2^2/\sigma_2^2}$$

Under the null hypothesis H_0: $\sigma_1^2 = \sigma_2^2$, the σ_1^2 and σ_2^2 can be canceled and the F statistic becomes

$$F = \frac{s_1^2}{s_2^2}$$

Because the areas less than .95 for F are not given in many tables, the usual procedure is to reverse the subscripts on the populations if necessary so that $s_1^2 \geq s_2^2$. Hence, we will make $s_1^2 = 1.9898$ and $s_2^2 = 1.0167$, and $n_1 = 12$ and $n_2 = 19$. The computed F value is

$$F = \frac{1.9898}{1.0167} = 1.96$$

In entering the F tables, the d.f. for the numerator are $n_1 - 1 = 11$ and the d.f. for the denominator are $n_2 - 1 = 18$. Since there is a different F distribution for every pair of sample sizes, F tables become quite extensive. In Table A.5, just four areas under the curve have been given, .025, .95, .975, and .99. Degrees of freedom for the numerator are denoted in the table by ν_1 , for the denominator by ν_2. The column headings of Table A.5 give ν_1, ν_2 is listed by rows.

For $\alpha = .05$ and a two-sided test, we will use the area of .975 (or the 97.5 percentile) of the F table. We do not have to check the lower percentile since we placed the larger variance in the numerator. In Table A.5, there are entries for $\nu_1 = 11$, but there are no entries for $\nu_2 = 18$. When there are no entries for a particular d.f., a conservative approach is to use the F value from Table A.5 for the closest lower d.f. For example, for $\nu_2 = 18$ the results tabled for $\nu_2 = 15$ can be used. For $\nu_1 = 11$ and $\nu_2 = 15$, the 97.5 percentile is given as $F = 3.01$. The correct value for the tabled F must lie between $F = 3.01$ for 15 d.f. and $F = 2.72$ for 20 d.f. Since our computed $F = 1.96$ is less than the correct value of F, we will not be able to reject the null hypothesis of equal variances.

If our computed F had been equal to 2.80, then we might want to find the percentile for $\nu_1 = 11$ and $\nu_2 = 18$, since by looking at Table A.5 it would not be clear whether we could accept or reject the null hypothesis. In this case, we could either perform the test using a statistical program that gave the actual P value or interpolate in Table A.5.

As a second example of the use of the F test, suppose that, with the data from two samples in Table 6.1, we wish to prove the assertion that the observed gains in weight under the standard diet are less variable than under the supplemented diet. Now, we will make a one-sided test of the hypothesis $\sigma_1^2 \leq \sigma_2^2$. The variance for the supplemented diet is $s_1^2 = 20{,}392$ and the variance for the standard diet is $s_2^2 = 7060$. We will reject the null hypothesis only if s_1^2 is very large compared to s_2^2. It clearly is larger, but it is not clear if it is enough larger without performing the test.

Under the null hypothesis, the computed F statistic is

$$F = \frac{s_1^2}{s_2^2} = \frac{20{,}392}{7060} = 2.89$$

The two sample sizes are $n_1 = 16$ and $n_2 = 9$, and so Table A.5 is entered with $\nu_1 = 15$ d.f. for the numerator and $\nu_2 = 8$ d.f. for the denominator. To test the null hypothesis with $\alpha = .05$, then we will compare the computed F to the 95 percentile in the F table since a one-sided test is being performed and the entire rejection region is in the right tail. The 95 percentile from Table A.5 is 3.22 so that we cannot reject the null hypothesis; we cannot conclude that the population variance for the supplemented diet is larger than that under the standard diet.

Note that in most cases a two-sided F test is performed since often the researcher does not know ahead of time which population is likely to have the larger variance. As was the case with computing confidence intervals in Section 10.2, it is important to screen the data for outliers before testing that two variances are equal using the F test.

10.4 APPROXIMATE *t* TEST

In the first F test of the hemoglobin levels of children with cyanotic and acyanotic heart disease, we were not able to reject the null hypothesis of equal variances. Similarly, we were not able to reject the null hypothesis in the diet example. Thus, in both cases it was possible that the population variances were equal. In Section 6.5.3, when we computed the confidence interval for the difference in the mean between two independent groups using the t distribution, we stated that one of the assumptions was equal population variances in the two groups. The same assumption was made in Section 7.2.3 when we tested whether the means of two independent populations were equal. The F tests that we performed in this chapter give us additional confidence that the t test of Section 7.2.3 was appropriate since we did not find significant differences in the variances.

But what if the variances had been significantly different? If the pooled-variance t test given in Chapter 7 is used when the population variances are not equal, the P value from the test may be untrustworthy. The amount of the error depends on how unequal the variances are and also on how unequal the sample sizes are. For nearly equal sample sizes, the error tends to be small.

One course of action that is commonly taken when we wish to perform a test concerning the means of two independent groups is to use an approximate t test. If the observations are independent simple random samples from populations that are normally distributed, then we can use the following test statistic:

$$t = \frac{\overline{X}_1 - \overline{X}_2}{\sqrt{s_1^2/n_1 + s_2^2/n_2}}$$

which has an approximate t distribution. Suppose in the hemoglobin example that $s_1^2 = 1.9898$, but that now $s_2^2 = 0.50$. If we perform the F test, it would be rejected at the $\alpha = .05$ level. Then, our computed approximate t statistic is

$$t = \frac{13.03 - 15.74}{\sqrt{1.9898/12 + 0.5/19}} = \frac{-2.71}{\sqrt{.1921}} = -6.18$$

The d.f. are approximately

$$\text{d.f.} = \frac{(s_1^2/n_1 + s_2^2/n_2)^2}{(s_1^2/n_1)^2/(n_1 - 1) + (s_2^2/n_2)^2/(n_2 - 1)}$$

We compute $s_1^2/n_1 = 0.1658$ and $s_2^2/n_2 = 0.0263$ and enter them into the formula for the d.f.,

$$\text{d.f.} = \frac{(0.1658 + 0.0263)^2}{(0.1658)^2/(12 - 1) + (0.0263)^2/(19 - 1)} = \frac{.0369}{.002499 + .000038} = 14.5$$

or, rounding up, 15 d.f. In general, the d.f. for the approximate test are less than for the usual t test. When the sample sizes are large, this makes little difference since the d.f. will be large enough so that a reduction in the d.f. does not appreciably affect the size of the tabled t value. For instance, looking at Table A.3 for the 95th percentile, if we reduce the d.f. by 5 for a small sample, from say 10 to 5, the tabled t value increases from 1.812 to 2.015. But if we reduce the d.f. for a larger sample by 5, from 55 to 50, the tabled t only increases from 1.673 to 1.676.

Statistical programs commonly include an option for using the approximate t test. The test statistic is the same among the various programs; however, slightly differing formulas are used for estimating the d.f. Some of these formulas are more conservative than others in that they result in fewer d.f.

Other options for handling unequal variances include checking that the two distributions from the samples appear to be from normal populations and trying transformations on the data before performing the t test. Sometimes the difference in variances is associated with a lack of normality so that the transformations discussed in Section 5.5 should be considered.

10.5 OTHER TESTS

When the variances are unequal, other tests may be considered. Sometimes the data are not normally distributed and finding a suitable transformation is impossible so that there are additional reasons for using a different test other than the usual t test. For example, a distribution all of whose observations are positive values and whose mode is zero cannot be transformed to a normal distribution using power transformations.

Sometimes the data are not measured using an equal interval scale (see Section 4.5.3) so that the use of means and standard deviations is questionable. Often psychological or medical data such as health status are measured using an ordinal scale.

Distribution-free methods can be used with data that do not meet the assumptions for the tests or confidence limits described in Chapters 6, 7, and in this chapter. Such tests used are commonly called *nonparametric* tests. Nonparametric tests are available that are suitable for single sample, independent groups or paired groups. For single samples or paired samples, either the sign test or the Wilcoxon signed-rank test are often used. For two independent samples, the Mann–Whitney (Wilcoxon) rank sum test is commonly used.

These tests are widely available in statistical programs. In this text, we do not cover these nonparametric tests but give three references that provide readable descriptions (see Conover [1999], Fisher and van Belle [1993], or Gibbons [1993].)

PROBLEMS

10.1 In Problem 6.3, the standard deviation of a population of bleeding times was given as 0.588 min, and it was assumed in estimating the mean bleeding time that when pressure is applied, the standard deviation of bleeding times remains the same. Use the sample of six bleeding times under pressure given in that problem (1.15, 1.75, 1.32, 1.28, 1.39, and 2.50 min) to obtain a 95% confidence interval for σ, the standard deviation of the population of bleeding times under pressure. Use the 95% confidence interval to test $H_0 : \sigma = 0.588$ min.

10.2 Use Problem 6.8 to make a test to decide whether the two variances are equal. Use a confidence level of .99.

10.3 In Problem 6.5, use the standard deviations for bleeding times calculated from 43 bleeding times without pressure and from 39 bleeding times under pressure to decide whether bleeding time under pressure is more variable than bleeding time without pressure.

10.4 From Table 7.2, make a 99% confidence interval for the variance of hemoglobin level, assuming that the variances are the same for acyanotic and cyanotic children, so that s_p^2 may be used.

10.5 In Problem 6.4, there was a sample of 12 litters of mice, each consisting of 1 male and 1 female, and the data consisted of 12 pairs of weights. Why is it inappropriate to use Table A.5, the F table, on these data if we wish to find out whether or not the variability of weights is the same among male and female mice?

10.6 Use the data in Table 6.2 to give a 99% confidence interval for the variance of the population of differences in weight gains.

10.7 Systolic blood pressure has been measured in two groups of older people who were not taking medication for high blood pressure. The first group of $n_1 = 21$ persons were overweight. They had a mean systolic blood pressure of $\overline{X}_1 = 142$ and $s_1 = 26$. The second group of $n_2 = 32$ was normal weight. They had $\overline{X}_2 = 125$ and $s_2 = 15$. Test the null hypothesis that $\sigma_1^2 = \sigma_2^2$ at the $\alpha = .05$ level. Then, test the null hypothesis of equal population means using the approximate t test.

REFERENCES

Conover, W. J. [1999]. *Practical Nonparametric Statistics*, 3rd ed., New York: Wiley.

Fisher, L. D. and van Belle, G. [1993]. *Biostatistics: A Methodology for the Health Sciences,* New York: Wiley-Interscience, 304–326.

Gibbons, J. D. [1993]. *Nonparametric Statistics An Introduction*, Newbury Park, CA: Sage Publications.

CHAPTER ELEVEN

Regression and Correlation

In previous chapters, one variable was usually studied at a time. An exception to this was Chapter 9, in which the odds ratio and the chi-square test were introduced for analysis of two categorical variables. In Chapter 9, we discussed analysis of categorical data from a single sample, from two or more samples, and from matched samples. In this chapter, a similar format is followed but here the data will be assumed to be continuous (either ratio or interval as described in Section 4.5.3).

First, we discuss the relationship between two variables when both variables are measured from a single sample. This is by far the most common use of the techniques given in this chapter. For example, we might take a sample of fifth graders and measure their height and weight. Or with a sample of adult males we might measure their age, height, weight, systolic blood pressure, and diastolic blood pressure. From this set of five variables, comparisons could be made two variables at a time. We shall also briefly mention the fixed-X case where the values of X are fixed in advance.

Usually, two variables are studied together in the general hope of determining whether there is some underlying *relation* between them, and if so, what kind of relationship it is. Sometimes, on the other hand, two variables are studied in the hope of being able to use one of them to *predict* the other. For example, we might have two methods of measuring a constituent of blood, one inexpensive to perform and the other time consuming and expensive. It would be useful to have a regression equation that would allow us to predict the more expensive result from the inexpensive one.

In Section 11.1, we discuss drawing and interpreting scatter diagrams, a widely used graphic method. Section 11.2 presents linear regression analysis when the observations are from a single sample. Formulas for computing the regression line, confidence intervals, and tests of hypotheses are covered. In Section 11.3, the correlation coefficient is defined. Confidence intervals, tests of hypotheses, and interpretation of the correlation coefficient are discussed. Section 11.4 discusses regression analysis for the fixed-X model: when the model is used and what can be estimated from this model. In Section 11.5, we discuss the use of transformations in regression analysis and the detection and effect of outliers.

11.1 THE SCATTER DIAGRAM: SINGLE SAMPLE

The simplest and yet probably the most useful graphical technique for displaying the relation between two variables is the scatter diagram (also called a scatter plot). The first step in making a scatter diagram is to decide which variable to call the *outcome* variable (also called the dependent or response variable) and which variable to call the *predictor* or independent variable. As the names imply, the predictor variable is the variable that we think predicts the outcome variable (the outcome variable is dependent on the predictor variable). For example, for children we would assume that age predicts height so that age would be the predictor variable and height the outcome variable—not the other way around.

The predictor variable is called the X variable and is plotted on the horizontal or X axis of the scatter diagram. The outcome variable is called the Y variable and is depicted on the vertical or Y axis. Each point on the scatter diagram must have both an X value and a Y value and is plotted on the diagram at the appropriate horizontal and vertical distances. As a small example of a scatter diagram, we will use the hypothetical data in Table 11.1, consisting of weights in pounds (lb) from a sample of 10 adult men as the predictor or X variable and their systolic blood pressure (SBP) in millimeters of mercury (mmHg) as the outcome or Y variable. The pair of values for each point is written as (X, Y). For example, in Table 11.1 the pair of values for the first adult male would be written as (165,134).

There are 10 points in the scatter diagram, one for each male. Scales have been chosen for the X and Y axes that include the range of the weights and of the systolic blood pressure. Tic marks have been placed at intervals of 5 units of systolic blood pressure and every 20 lb of weight. Most statistical programs will do this automatically. Some programs also allow the user to select the end points and the intervals. This is a useful option if one wishes to make a scatter diagram with the same end points and intervals as a previous scatter diagram.

Table 11.1. Weights and Systolic Blood Pressure for 10 Males

Number	Weight (lb)	SBP (mmHg)
1	165	134
2	243	155
3	180	137
4	152	124
5	163	128
6	210	131
7	203	143
8	195	136
9	178	127
10	218	146

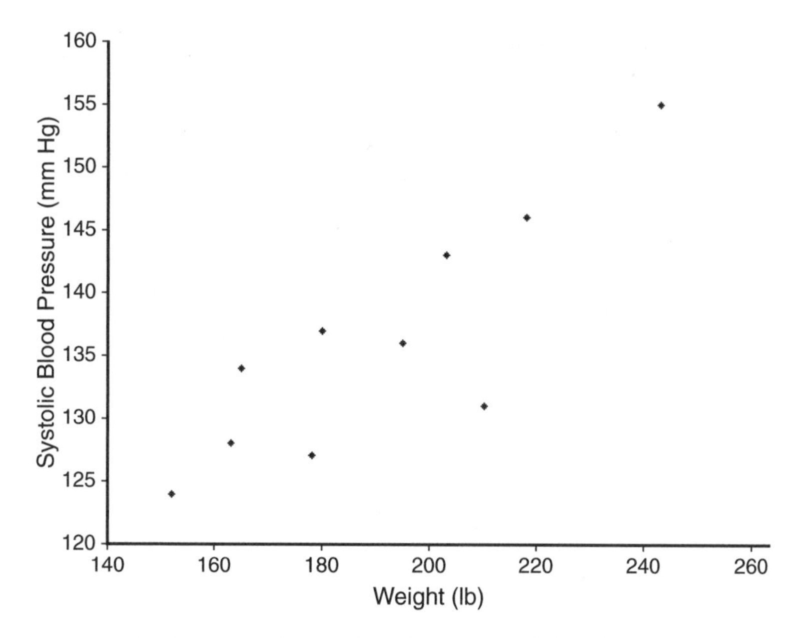

Fig. 11.1. Scatter diagram of weight versus systolic blood pressure.

The scatter diagram is extremely useful in indicating the relationship between the predictor and outcome variables (see Fig. 11.1). One thing we note is whether the relationship between X and Y is positive or negative. In Figure 11.1, it can be seen that systolic blood pressure increases as weight increases. Adult males in this sample who have higher weight tend to have higher blood pressure. This is called a positive relationship. If we had plotted data from adults using vital capacity as the outcome variable and age as the predictor variable, we would have a negative relationship since vital capacity tends to decrease with increasing age.

If the points fall roughly within a circle, there is essentially little or no appreciable relationship between the predictor variable and the outcome variable. If all the points fall close to a straight line or curve, then we say there is a strong relationship between X and Y. The points in Figure 11.1 tend to fall fairly close to a straight line; in this sample systolic blood pressure tends to increase linearly with weight.

The points on many scatter diagrams seem to follow a straight line at least over a limited range of the X variable. However, in some examples the points appear to follow a curve. Additional methods for examining the relationship between X and Y from scatter diagrams are given in Chambers et al. [1983] and Cleveland [1985].

11.2 LINEAR REGRESSION: SINGLE SAMPLE

In this section, we first show how to compute a linear regression line, and then how to interpret it.

11.2.1 Least-Squares Regression Line

After plotting the scatter diagram, we would like to fit a straight line or a curve to the data points. Fitting curves will not be discussed here, except that later we will show how a transformation on one of the variables will sometimes enable us to use a straight line in fitting data that was originally curved.

A straight line is the simplest to fit and is commonly used. Sometimes a researcher simply draws a straight line by eye. The difficulty with this approach is that no two people would draw the same line. We would like to obtain a line that is both *best* in some sense and that also is the same line that other investigators use. The line with both these attributes is the *least-squares regression line*.

The equation of the line is

$$\hat{Y} = a + bX$$

Here, $\hat{Y}$ denotes the value of Y on the regression line for a given X. The coordinates of any point on the line are given as $(\hat{Y}, X)$. The slope of the line is denoted by b and the intercept by a. The numerical value of b can be calculated using the formula

$$b = \frac{\sum(X - \overline{X})(Y - \overline{Y})}{\sum(X - \overline{X})^2}$$

and the numerical value of a can be obtained from

$$a = \overline{Y} - b\overline{X}$$

Before the interpretation of the regression line is discussed, the example given in Table 11.1 will be used to demonstrate the calculations. Note that almost all statistical programs will perform these calculations so this is for illustration purposes. The calculation of a regression line for a large sample is quite tedious as is obvious from Table 11.2. We first calculate the mean weight as $\overline{X} = 1907/10 = 190.7$ and the mean systolic blood pressure as $\overline{Y} = 1361/10 = 136.1$. Then, for the first row in Table 11.2, we obtain $(X - \overline{X})^2$ by calculating $(X - 190.7)$ or $(165 - 190.7)$ and squaring the difference of -25.7 to obtain 660.49. A similar calculation is done for $(Y - 136.1)^2$ to obtain 4.41. The value in the first row and last column is computed from $(X - 190.7)(Y - 136.1) = (-25.7)(-2.1) = 53.97$. The last three columns of Table 11.2 are filled in a similar fashion for rows numbered 2 through 10 and the summation now is computed.

From the summation row, we obtain the results we need to compute

$$b = \frac{\sum(X - \overline{X})(Y - \overline{Y})}{\sum(X - \overline{X})^2} = \frac{2097.3}{7224.1} = .2903$$

and

$$a = \overline{Y} - b\overline{X} = 136.1 - .2903(190.7) = 80.74.$$

Table 11.2. Calculations for Regression Line

Number	Weight	SBP[a]	$(X-\overline{X})^2$	$(Y-\overline{Y})^2$	$(X-\overline{X})(Y-\overline{Y})$
1	165	134	660.49	4.41	53.49
2	243	155	2735.29	357.21	988.47
3	180	137	114.49	0.81	−9.63
4	152	124	1497.69	146.41	468.27
5	163	128	767.29	65.61	224.37
6	210	131	372.49	26.01	−98.43
7	203	143	151.29	47.61	84.87
8	195	136	18.49	0.01	−0.43
9	178	127	161.29	82.81	115.57
10	218	146	745.29	98.01	270.27
$\sum$	1907	1361	7224.10	828.90	2097.30

[a] Systolic blood pressure (SBP).

Then, $\hat{Y}$ equals

$$\hat{Y} = 80.74 + .2903X$$

or

$$\text{SBP} = 80.74 + .2903 \text{ weight}$$

Note that if we divide $\sum(X-\overline{X})^2$ by $n-1$, we obtain the variance of X or s_X^2. Similarly dividing $\sum(Y-\overline{Y})^2$ by $n-1$, or in this example by 9, we get the variance of Y or s_Y^2. Thus, the sum in those two columns is simply the variance of X multiplied by $n-1$ and the variance of Y multiplied by $n-1$. These sums are always positive since they are the sum of squared numbers. The sum of the last column in Table 11.2 introduces a formula that is new, $\sum(X-\overline{X})(Y-\overline{Y})$. If we divide this sum by $n-1=9$, we obtain what is called the *covariance* or s_{xy}. The covariance will be positive if increasing values of X result in increasing values of Y (a positive relationship holds). In Table 11.2, we have a positive sum of 2097.3. Note that only three of the values in, the last column are negative and two of those (−9.63 and −0.43) are small numbers. In other words, when X is $> \overline{X}$, then Y also tends to be $> \overline{Y}$ resulting in a positive product in this example.

Whenever large values of X tend to occur with small values of Y, then $\sum(X-\overline{X})(Y-\overline{Y})$ is negative.

In the formula for the slope b, we divide $\sum(X-\overline{X})(Y-\overline{Y})$ by a positive number; thus the resulting sign of b depends solely on the sign of $\sum(X-\overline{X})(Y-\overline{Y})$. If it has a positive sign, then b has a positive sign and the relationship is called positive. If it has a negative sign, then b has a negative sign and the relationship is a negative one.

11.2.2 Interpreting the Regression Coefficients

The quantity $b = .2903$ is called the *slope* of the straight line. In the case of the regression line, b is called the *regression coefficient*. This number is the change in $\hat{Y}$ for a unit change in X. If X is increased by 1 lb, $\hat{Y}$ is increased by .2903 mmHg. If X is increased by 20 lbs, $\hat{Y}$ is increased by $.2903(20) = 5.8$ mmHg. Note that the slope is positive and so a heavier weight tends to be associated with a higher systolic blood pressure. If the slope coefficient were negative, then increasing values of X would tend to result in decreasing values of Y. If the slope were 0, then the regression line would be horizontal.

One difficulty in interpreting the value of the slope coefficient b is that it changes if we change the units of X. For example, if X were measured in kilograms instead of pounds we would get a smaller value for the slope. Thus it is not obvious how to evaluate the magnitude of the slope coefficient. One way of evaluating the slope coefficient is to multiply b by $\overline{X}$ and to contrast this result with $\overline{Y}$. If $b\overline{X}$ is small relative to $\overline{Y}$, then the magnitude of the effect of b in predicting Y will tend to be small.

The quantity $a = 80.74$ is called the *intercept*. It represents the value of $\hat{Y}$ when $X = 0$. The magnitude of a is often difficult to interpret since in many regression lines we do not have any values of X close to 0, and it is hard to know if the points fit a straight line outside of the range of the actual X values. Since no adult male could have a weight of 0, the value of the intercept is not very useful in our example.

11.2.3 Plotting the Regression Line

The regression line may be plotted as any straight line is plotted, by calculating the value of $\hat{Y}$ for several values of X. The values of X should be chosen spaced sufficiently far apart so that small inaccuracies in plotting will not influence the placement of the line too much. For example, for the regression line,

$$\hat{Y} = 80.74 + .2903X$$

we can substitute $X = 140$, $X =$ the mean 190.7, and $X = 260$ in the equation for the regression line and obtain $\hat{Y} = 121.4$, $\hat{Y} = 136.1$, and $\hat{Y} = 156.2$. The 10 original points are shown in Figure 11.2 as well as the regression line. Note that the regression line has only been drawn to include the range of the X values. We do not know how a straight line might fit the data outside the range of the X values that we have measured and plotted.

Note also that at $\overline{X} = 190.7$, the height of the line is $\hat{Y} = 136.1 = \overline{Y}$. This is always the case. The least-squares regression line always goes through the point $(\overline{X}, \overline{Y})$.

11.2.4 The Meaning of the Least Squares Line

If for each value of X, we calculate $\hat{Y}$, the point on the regression line for that X value, then it can be subtracted from the observed Y value to obtain $Y - \hat{Y}$. The difference

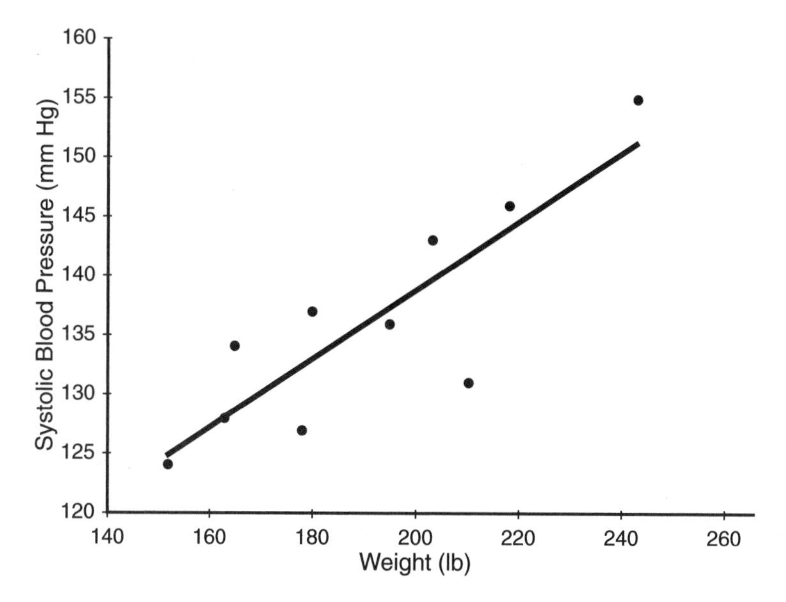

Fig. 11.2. Least-squares regression line, $\hat{Y} = 80.74 + .2903X$.

$Y - \hat{Y}$ is called a *residual*. The residuals are the vertical distances of data points from the regression line. For the example with the 10 adult males, we would have 10 residuals. Each residual indicates whether the systolic blood pressure is high or low considering the individual's weight.

Table 11.3 illustrates the calculation of the residuals and their squares, which are needed in Section 11.2.5. The predicted $\hat{Y}$ is computed for each value of X. For example, for the first X of 165, we have $\hat{Y} = 80.74 + .2903(165) = 128.64$. The

Table 11.3. Calculations of Residuals

Number	X	Y	$\hat{Y}$	$(Y - \hat{Y})$	$(Y - \hat{Y})^2$
1	165	134	128.64	5.36	28.73
2	243	155	151.28	3.72	13.84
3	180	137	132.99	4.01	16.08
4	152	124	124.86	−0.86	0.74
5	163	128	128.06	−0.06	0.00
6	210	131	141.70	−10.70	114.49
7	203	143	139.67	3.33	11.09
8	195	136	137.35	−1.35	1.82
9	178	127	132.41	−5.41	29.27
10	218	146	144.03	1.97	3.88
$\sum$	1907	1361	1361	0.01	219.94

residual is then calculated as $134 - 128.64 = 5.36$. When we square the residual we get $(5.36)^2 = 28.73$. The other rows follow in a similar fashion.

If the 10 values of $Y - \hat{Y}$ are squared and added together, as has been done in Table 11.3, the sum $\sum(Y - \hat{Y})^2$ is calculated to be 219.94. This sum is smaller for the line $\hat{Y} = 80.74 + .29X$ than it would be for any other straight line we could draw to fit the data. For this reason, we call the line the *least-squares line*. The regression line is considered to be the best-fitting line to the data in the sense that the sum of squares in the *vertical* direction is as small as possible.

Note that in Table 11.3 the sum of the 10 values of $Y - \hat{Y}$ is approximately 0. If there were no rounding off in the calculations, it would be precisely 0. The sum of vertical distances from the regression line is always 0. Also, the sum of the $\hat{Y}$'s always equals the sum of Y's (within rounding error).

11.2.5 The Variance of the Residuals

The 10 residuals $Y - \hat{Y}$, as calculated in Table 11.3, are simply a set of 10 numbers, and so we can calculate their variance. This variance is called the *residual mean square* and its formula is

$$s_{y.x}^2 = \frac{\sum(Y - \hat{Y})^2}{n - 2}$$

For the 10 data points, $s_{y.x}^2 = 219.94/(10-2) = 27.49$, and taking the square root we obtain $s_{y.x} = 5.24$ mmHg for the standard deviation of the residuals. An obvious difference between the variance of the residuals and the variances of X and Y that we obtained earlier is that here we divided by $n - 2$ instead of $n - 1$. The $n - 2$ is used because both the slope and the intercept are used in computing the line. The square root of the residual mean square is often called the *standard error of estimate*.

The residual mean square measures the variation of the Y observations around the *regression line*. If it is large, the vertical distances from the line are large. If it is small, the vertical distances are small. In the example using the 10 data points, $s_{y.x}^2 = 27.49$. The variance of the original 10 Y values can be computed from Table 11.2 as $828.90/9 = 92.1$, so the variance about the regression line is much smaller than the variance of Y in this example. This indicates that Y values tend to be closer to $\hat{Y}$ than they are to $\overline{Y}$. Thus, using the least-squares regression line to predict systolic blood pressure from weight gives us a closer prediction than using the sample mean systolic blood pressure.

11.2.6 Model Underlying Single Sample Linear Regression

Up to this point in this chapter, we have discussed computing a regression line from a single sample. To compute confidence limits and to make tests of hypotheses, we shall make the basic assumption that we have taken a simple random sample from a population where X and Y follow a bivariate normal distribution. In Chapter 5, we discussed a normal distribution for a single variable X. For X and Y to be bivariately

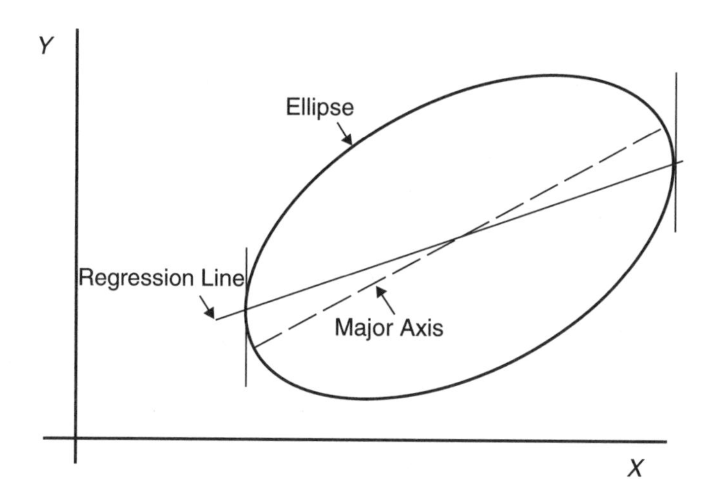

Fig. 11.3. Position of regression line for bivariate normal distribution.

normal requires not only that both X and Y be individually normally distributed but also that the relationship between X and Y has a bivariate normal distribution (see Fisher and van Belle [1993]).

One way of determining whether the data points may be from a bivariate normal distribution is to examine the scatter diagram. If the variables have a bivariate normal distribution, data points should lie approximately within an ellipse. An ellipse is depicted in Figure 11.3. Also, if the data points (X, Y) follow a bivariate normal distribution, then both X and Y separately should be normally distributed. Normal probability plots can be used to check this. The residuals should be normally distributed; many computer programs will allow the user to check the distribution of the residuals using normal probability plots or other visual methods.

If we enclose the points in Figure 11.1 with an ellipse it would be a long and thin ellipse, thus indicating a strong relationship between X and Y. An ellipse closer to a circle would be an indication of a weak relationship between X and Y.

Figure 11.3 also shows the position of the least-squares regression line. Note that in Figure 11.3 the regression line is not the major (or principal) axis of the ellipse but tends to have a smaller slope than the major axis. This is the case whenever the relationship is positive. If the relationship is negative (negative slope coefficient), then the slope of the regression line will be closer to 0 than the major axis of the ellipse. In both cases, the regression line is more nearly horizontal than the major axis of the ellipse. The least-squares line goes through the center of the ellipse and touches the ellipse at the points where the two vertical lines touch the ellipse (are tangent to the ellipse).

As an aside, the fact that the slope coefficient will be smaller than that of the major axis of the ellipse when the relationship is positive has special importance in interpreting data taken from paired samples. For example, suppose that cholesterol levels were taken prior to treatment and 1-month later. We could make a scatter plot of the data with the pretreatment cholesterol level on the X axis and the posttreatment level on the Y axis. If there was no change except for that due to measurement error and day-to-day variation in levels at the two time periods, then we would expect to see the points in a scatter diagram falling roughly within an ellipse with the major axis having a 45° slope (slope of 1) and going through the origin. The pretreatment and posttreatment means would be roughly equal. Note that since we will fit a least-squares line, the regression line will likely have a slope < 1, possibly quite a bit less if there is a great deal of variation in the results. If all the points fall very close to a straight line, then there is very little difference between the slope of the major axis and that obtained from the least-squares regression line. Note that the same considerations come into play if we wished to compare the readings made by two different observers or compared two different laboratory methods for measuring some constituent of blood.

If the assumption of a simple random sample from a bivariate normal population can be made, then we can use our sample statistics to estimate population parameters. The population parameters that can be estimated and the sample statistics used to estimate them are given in Table 11.4. Note from Table 11.4 that α and β are used in linear regression to identify the intercept and slope coefficient for the population regression line. This follows the convention of using Greek letters for population parameters. In Section 7.4, α was used to denote the chance of making a Type I error and β was used to denote the chance of making a Type II error. To avoid confusion in this chapter, we always say population intercept and population slope when referring to the population regression line.

In earlier chapters, we have already covered the estimation of μ_X, μ_Y, σ_x^2, and σ_y^2 when variables are considered one at a time. The following sections discuss confidence intervals and tests of hypotheses concerning the remaining estimators.

11.2.7 Confidence Intervals in Single Sample Linear Regression

If the required assumptions stated in the previous Section 11.2.6 are made, confidence intervals can be made for any of the parameters given in Table 11.4. The most

Table 11.4. Parameters and Statistics for Single Sample Case

Population Parameters	Sample Statistics	Description
μ_X	$\overline{X}$	Mean of X
μ_Y	$\overline{Y}$	Mean of Y
σ_x^2	s_x^2	Variance of X
σ_y^2	s_y^2	Variance of Y
$\sigma_{y.x}^2$	$s_{y.x}^2$	Variance of the residuals
β	b	Slope of line
α	a	Intercept of line
ρ	r	Correlation coefficient

widely used confidence interval is the one for the slope of the regression line. When computing confidence intervals for a mean in Chapter 6, we used the sample mean plus or minus a t value from Table A.3 times the standard error of the mean. We will follow the same procedure in computing confidence intervals for b, the slope of the line. The standard error of b is given by

$$\text{se}(b) = \frac{s_{y.x}}{\sqrt{\sum(X - \overline{X})^2}}$$

Note that in computing se(b), we divide the standard error of the estimate, $s_{y.x}$, by the square root of $\sum(X - \overline{X})^2$. Hence, the se(b) becomes smaller as the sample size increases. Further, the more spread out X's are around their mean, the smaller se(b) becomes.

The 95% confidence interval is given by

$$b \pm t[.975][\text{se}(b)]$$

For example, for the 10 males we have already computed the square root of the residual mean-square error or $s_{y.x} = 5.24$. From Table 11.2, we have $\sum(X - \overline{X})^2 = 7224.1$. So the se$(b)$ is

$$se(b) = \frac{5.24}{\sqrt{7224.1}} = \frac{5.24}{84.99} = .062$$

The 95% confidence interval is

$$.29 \pm 2.306(.062) = .29 \pm .14$$

That is, the confidence interval for the population slope is

$$.15 < \beta < .43$$

Here we used a t value from Table A.3, which corresponds to the area up to .975 in order to obtain 95% two-sided intervals. The d.f. are $n - 2 = 10 - 2 = 8$. The $n - 2$ must be used here because it was necessary to estimate two population parameters of the regression line, α and β. If we repeatedly took samples of size 10 and calculated the 95% confidence intervals, 95% of the intervals would include β, the population slope coefficient.

Occasionally, we need to compute confidence intervals for the population intercept α. If the sample includes points that have X values close to 0 so that the regression line does not extend far below the smallest X value, then these confidence intervals can safely be interpreted. The standard error of a is given by

$$\text{se}(a) = s_{y.x}\left[\frac{1}{n} + \frac{\overline{X}^2}{\sum(X - \overline{X})^2}\right]^{1/2}$$

and the 95% confidence interval for the population intercept is

$$a \pm t[.975][\text{se}(a)]$$

where a t value with $n - 2$ d.f. is used. For the example with the 10 males, we have

$$\text{se}(a) = 5.24[1/10 + (190.7)^2/7224.1]^{1/2} = 5.24(2.266) = 11.87$$

and the 95% confidence interval is computed from

$$80.74 \pm 2.306(11.87) = 80.74 \pm 27.37$$

or

$$53.37 < \alpha < 108.11$$

Note that for this example, there is no reason to interpret the confidence limits for the population intercept; no weights are near 0. In general, caution should be used in making inferences from the regression line below the smallest X value or above the largest X value.

The confidence interval for the variance about the regression line follows the same procedure used in Chapter 10 for a single sample except that $n - 1$ is replaced by $n - 2$. The values for chi-squared are taken from Table A.4 for 8 d.f. The formula is given by

$$\frac{(n-2)s_{y.x}^2}{\chi^2[.975]} < \sigma_{y.x}^2 < \frac{(n-2)s_{y.x}^2}{\chi^2[.025]}$$

and entering the data for the 10 males, we have

$$\frac{8(27.49)}{17.53} < \sigma_{y.x}^2 < \frac{8(27.49)}{2.18}$$

or $12.55 < \sigma_{y.x}^2 < 100.88$. Taking the square root, we have $3.54 < \sigma_{y.x} < 10.04$ for the 95% confidence interval of the standard deviation of the residuals about the regression line.

Most computer programs furnish only limited or no confidence intervals for linear regression. They typically give the standard error of the regression line, the slope coefficient, and the intercept. The variance of X can be used to compute $\sum(X - \overline{X})^2$. From these quantities the desired confidence intervals can be easily computed.

11.2.8 Tests of Hypotheses for Regression Line from a Single Sample

The computation of tests of hypotheses for the population intercept and the slope are common options in computer programs. In Chapter 7, we presented the test to decide whether a population mean was a particular value. A similar test can be made for the slope of a regression line. The test statistic is

$$t = \frac{(b - \beta_0)}{\text{se}(b)}$$

Here, β_0 is the hypothesized population slope and the computed t is compared with the t values in Table A.3 with $n - 2$ d.f. The se(b) is given in Section 11.2.7. For example, if we wish to test the null hypothesis that the population slope, $\beta_0 = .20$, for the 10 males and with $\alpha = .05$, our test statistic would be

$$t = \frac{(.29 - .20)}{.062} = 1.45$$

From Table A.3 for $n - 2 = 8$ d.f., the tabled value of t is 2.306 and we would be unable to reject the null hypothesis of a population slope equal to .20.

Similarly, for a test that the population intercept α takes on some particular value, α_0, we have the test statistic

$$t = \frac{(a - \alpha_0)}{\text{se}(a)}$$

For example, to test the null hypothesis that the population intercept $\alpha_0 = 100$ mmHg, for the 10 males at the 5% level, we have

$$t = \frac{(80.74 - 100)}{11.87} = -1.62$$

and we would not be able to reject the null hypothesis since, for $n - 2 = 8$ d.f., we need a t value greater in magnitude than the tabled value of 2.306 in Table A.3 to reject at the $\alpha = .05$ significance level.

Statistical programs test the null hypotheses that the population slope $\beta_0 = 0$ and that population intercept $\alpha_0 = 0$. If the test that the population slope $\beta_0 = 0$ is not rejected, then we have no proof that Y changes with different values of X. It may be that $\overline{Y}$ provides just as good an estimate of the mean of the Y distribution for different values of X as does $\hat{Y}$. Tests of other hypothesized values must be done by hand. However, most programs furnish the standard errors so that these other tests can be made with very little effort.

The test that the population intercept $\alpha_0 = 0$ is a test that the line goes through the origin (0,0). There is usually no reason for making this test.

11.3 THE CORRELATION COEFFICIENT FOR THE SINGLE SAMPLE

When the relation between two variables from a single sample is studied, it often seems desirable to have some way of measuring the degree of association or correlation between them. For categorical data in Section 9.2.2, we presented the odds ratio as a measure of association. Here, for continuous data interval or ratio data, the most widely used measure of association is the correlation coefficient r.

11.3.1 Calculation of the Correlation Coefficient

The definition of the correlation coefficient is

$$r = \frac{\sum(X - \overline{X})(Y - \overline{Y})}{\sqrt{\sum(X - \overline{X})^2 \sum(Y - \overline{Y})^2}}$$

Before discussing the meaning of r, we shall illustrate its calculation from the calculations done for the regression line in Table 11.2. From Table 11.2, we have $\sum(X - \overline{X})(Y - \overline{Y}) = 2097.3$, $\sum(X - \overline{X})^2 = 7224.1$, and $\sum(Y - \overline{Y})^2 = 828.9$. Substituting these numerical values in the equation for r we have

$$r = \frac{2097.3}{\sqrt{7224.1(828.9)}}$$

or

$$r = \frac{2097.3}{\sqrt{5,988,056.5}} = \frac{2097.3}{2447.05} = .857$$

Thus, the correlation between weight and systolic blood pressure for the 10 adult males is .857.

11.3.2 The Meaning of the Correlation Coefficient

As was the case for the slope coefficient b, the sign of r is determined by its numerator, $\sum(X - \overline{X})(Y - \overline{Y})$, since the denominator is always positive. Thus, if values of X greater than $\overline{X}$ occur when values of Y are greater than $\overline{Y}$ and small values of X occur when small values of Y occur, then the value of the r will be positive. We then have a positive relationship. In the example with the 10 males, r is positive so that large values of weight are associated with large systolic blood pressure. If we had computed a negative r, then high values of Y would occur with low values of X.In the example of this given earlier, lower vital capacity tends to occur with older age.

It can be shown that r always lies between -1 and $+1$. Indeed, if all the data points lie precisely on a straight line with a negative slope, the correlation coefficient is always exactly -1. If all the points lie on a straight line with positive slope, then $r = +1$. Figure 11.4 illustrates these possibilities.

A correlation coefficient of 0 is interpreted as meaning that there is no linear relation between the two variables. Figure 11.5 illustrates data with zero correlations. Note that the correlation coefficient can be 0 and still there may possibly be a nonlinear

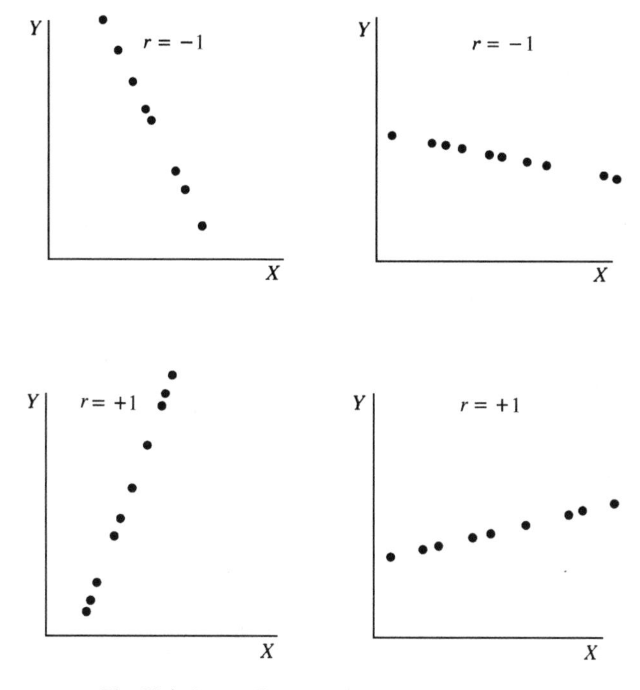

Fig. 11.4. Scatter diagrams with $r = -1$ and $r = +1$.

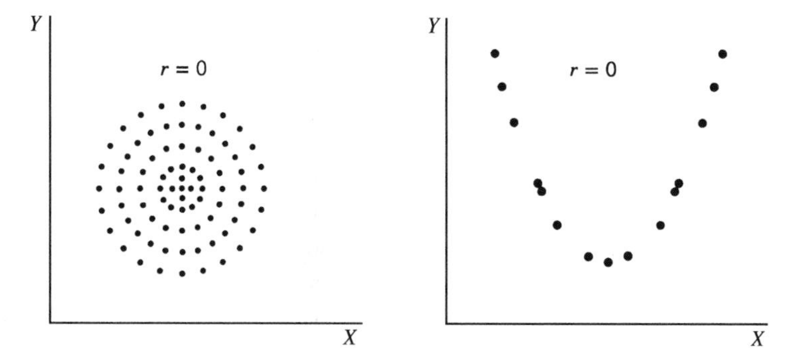

Fig. 11.5. Scatter diagrams with $r = 0$.

relation between the two variables. The correlation coefficient is a measure of the *linear* relationship between X and Y.

Examining the magnitude of r cannot replace a look at the scatter diagram, but it is a useful measure of the degree of linear relationship between two variables. A value of r less than $-.7$ or greater than $+.7$ in a large set of data might be considered to indicate a high degree of relationship. It can be seen from comparing the formulas for r and b that they always have the same sign. A negative value of r indicates that the regression line slopes downward. However, the size of r does not indicate the size of the slope coefficient.

11.3.3 The Population Correlation Coefficient

The observations are assumed to come from a single simple random sample of n individuals from a population with X and Y bivariately normally distributed. Each individual in the population has a value of X and a value of Y (weight and systolic blood pressure, for example). We can think of this population as having a correlation coefficient just like that of a sample. The population coefficient of correlation will be denoted by the Greek letter ρ, with r kept for the sample correlation coefficient.

Just as for other population parameters such as means and slopes, it is possible to obtain confidence intervals for the population ρ and to compute tests of hypotheses.

11.3.4 Confidence Intervals for the Correlation Coefficient

A correlation coefficient calculated from a small sample may be quite misleading, so that confidence intervals for the population correlation coefficient are useful. There are various methods for obtaining such confidence intervals; here only one graphical method will be given.

Table A.6 gives a chart from which 95% confidence intervals can be read with accuracy sufficient for most purposes. For the weight and systolic blood pressure example, $r = .857$. To find the 95% confidence interval, the point .857 is found on the horizontal scale at the bottom of the chart. We then look directly upward in a vertical line until we find a curve labeled 10, the sample size. We then read the value from the vertical scale on the left or on the right of the chart corresponding to the point on the curve labeled 10 just above .857. This value is approximately .50, which is the lower confidence limit. For the upper confidence limit, we find .857 on the horizontal scale across the top of the chart and look directly down to the upper curve labeled 10. Reading across then the approximate point .97 is found, which is the upper confidence limit. Thus the 95% confidence interval for ρ is approximately $.50 < \rho < .97$, a rather wide range. Because the lower and upper confidence limits are both positive, we conclude that the population correlation coefficient is positive.

If the sample size is not listed on the chart, conservative intervals can be obtained by using the next smaller sample size that is given on the chart. Alternatively, we can interpolate roughly between two of the curves. For example, if our sample size was 11, we could either use a sample size of 10 or interpolate half-way between 10 and 12 in Table A.6. More accurate estimates of the confidence intervals can be obtained using a method given in Dunn and Clark [1987].

The confidence intervals calculated in this section can be used to test $H_0 : \rho = \rho_0$, where ρ_0 is any specified value between -1 and $+1$. For the sample of size 10 with $r = .857$, we might wish to test $H_0 : \rho = .7$. Since .7 is contained in the 95% confidence interval +.50 to +.97, the null hypothesis is not rejected.

11.3.5 Test of Hypothesis That $\rho = 0$

The most commonly used test is of H_0: $\rho = 0$; in other words a test of no association between X and Y.We can make this test by using the test statistic

$$t = \frac{r}{\sqrt{(1 - r^2)/(n - 2)}}$$

By substituting in the values from the sample of 10 adult males we have

$$t = \frac{.857}{\sqrt{(1 - .857^2)/(10 - 2)}} = \frac{.857}{\sqrt{.26555/8}}$$

or

$$t = \frac{.857}{.1822} = 4.70$$

If we compare the computed t value with a t value of 2.306 with $n-2 = 8$ d.f. from Table A.3, we would reject the H_0: $\rho = 0$ at the $\alpha = .05$ significance level. Note that the test of $\rho = 0$ is equivalent to the test that the population slope $\beta = 0$. If we cannot say that the population slope is not 0, we cannot say that the population correlation is not 0. This test of H_0: $\rho = 0$ is commonly included in statistical programs but the test that ρ is some value other than 0 is not included.

11.3.6 Interpreting the Correlation Coefficient

An advantage held by the correlation coefficient over the slope of the regression line is that the correlation coefficient is unaffected by changes in the units of X or Y. In the example, if weight had been in kilograms, we would still obtain the same correlation coefficient, .857. In fact, r is unaffected by any linear change in X and Y. We can add or subtract constants from X or Y or multiply X or Y by a constant and the value of r remains the same.

The correlation r has a high magnitude when the ellipse depicted in Figure 11.3 is long and thin. All the points lie close to a straight line. A value of r close to 0 results if the points in a scatter diagram fall in a circle or the plot is nonlinear. Scatter diagrams are a great help in interpreting the numerical value of r.

When a high degree of correlation has been established between two variables, one is sometimes tempted to conclude that "a causal relation between the two variables has been statistically proved." This is simply not true, however. There may or may not be a cause-and-effect relationship; all that has been shown is the existence of a straight-line relationship.

An explanation of a high correlation must always be sought very carefully. If the correlation between X and Y is positive and high, then possibly a large X value may tend to make an individual's Y value large or perhaps both X and Y are affected by some other variables. The interpretation must be based on knowledge of the problem, and various possible interpretations must be considered.

11.4 LINEAR REGRESSION ASSUMING THE FIXED-X MODEL

Up to this point in Chapter 11, we have been discussing the single sample model. The single sample model is suitable in analyzing survey data or when examining data from one of two or more groups. We will now take up the fixed-X model. This is a briefer section since the formulas given for the single sample model remain the same.

11.4.1 Model Underlying the Fixed-X Linear Regression

In the fixed-X model, the values of X are selected by the investigator or occur because of the nature of the situation. For example, in studying the effects of pressure bandages on leg circumference, nurses were randomly assigned to wear a pressure bandage for either 30, 60, or 120 min. The outcome or Y variable is the difference in leg circumference before minus after applying the pressure bandage. In this experiment, the three time periods are the X variable. The values of X are considered to be fixed and known without error. This example would be a case where the values of X are selected by the investigator. This model is useful in analyzing experiments when X values (the variable whose values are set by the investigator) are continuous (interval or ratio data).

Alternatively, in studies of change over time, data are obtained on some outcome variable such as mortality rates or expected length of life by year. The year becomes the fixed-X variable. Usually, the investigator does not take a random sample of years but instead looks at the most recent 20 years or some other time period of interest. Here, the nature of the situation dictates how the sample is taken. Again, the X variable will be assumed to be fixed and known without error.

To make inferences to the population from confidence intervals or tests of hypotheses, different assumptions are made in this model than in the single sample case. For each value of X being considered, a population of Y values exists and the following three assumptions must hold:

1. The Y values at each X are normally distributed.
2. Their means lie on a straight line.
3. Their variances are equal to each other.

Figure 11.6 presents a hypothetical example where measurements have been taken at three values of X, that is, $X(1)$, $X(2)$, and $X(3)$. The Y values for each X value are normally distributed with the same variance and the mean of each set of Y values falls on a straight line.

11.4.2 Linear Regression Using the Fixed-X Model

The computation of the regression line, $\hat{Y} = a + bX$, proceeds exactly the same for the fixed-X model as the bivariate normal model. The same formulas are used for a, b, $s_{y.x}$, se(a), and se(b). The confidence intervals for the population intercept and

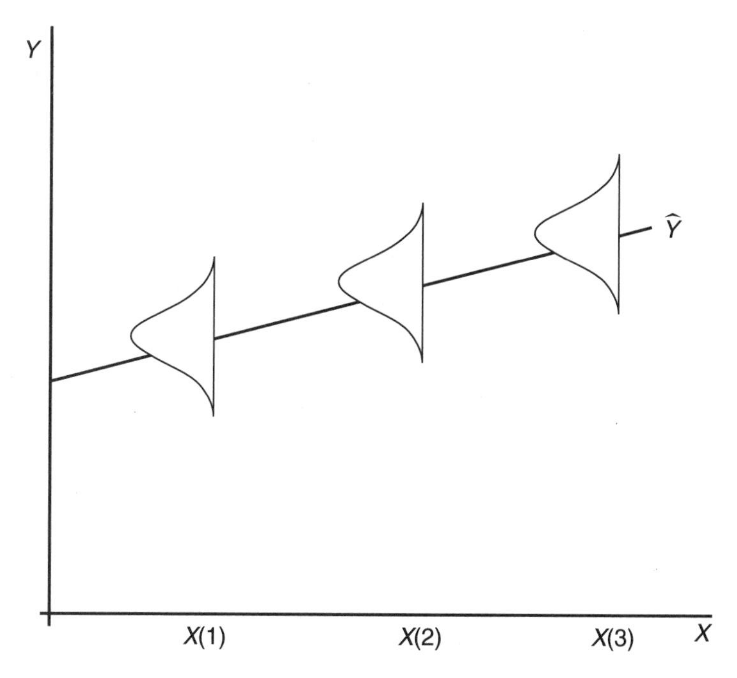

Fig. 11.6. Graph depicting assumptions for the fixed-X model.

slope, α and β are precisely the same. Tests of hypotheses concerning the population α and β are also the same as those given for the single sample bivariate normal mode.

With the fixed-X model we cannot estimate all the parameters that we could for the bivariate normal model. For the fixed-X model we cannot estimate the parameters μ_X, μ_Y, σ_x, σ_y, σ_{xy}, or ρ. The sample statistics for these parameters do not provide valid estimates of the population parameters. Note that computer programs will print out $\overline{X}$, $\overline{Y}$, s_x, s_y, and r, but these should be ignored if this model is assumed.

We lose the ability to estimate μ_X and σ_x because X values were deliberately chosen. For example, if we chose values of X that are far apart, we will get a larger variance and standard deviation than we would if we choose them close together. Since we are assuming that the Y values depend on X, they also are affected and we cannot estimate μ_Y or σ_y. The correlation r is also affected by the choice of X values, so we cannot estimate ρ.

One advantage of the fixed-X model is that when individuals are randomly assigned to say three treatment levels as given in the leg circumference example, we can make causal inferences. In the case of the mortality rates for the most recent years, direct causality cannot be established. If we find changes in the rates over time, we do not know what *caused* them.

The assumptions made in using the fixed-X model should first be checked graphically by looking at a scatter diagram with the line drawn on it. The points should fall along a straight line with the residuals from the line approximately the same size for different values of X (equal variance assumption). If there are multiple Y values at a limited number of X values, then the means of the Y values at each X value should fall close to the line and the range of the Y values for each X value should be similar. Indications of not meeting the assumptions would be to have the ranges of the Y values at each X value increase with increasing levels of the means of the Y variable or to have the means of Y for various values of X not fall close to a straight line.

11.5 OTHER TOPICS IN LINEAR REGRESSION

In this section, two methods of correcting potential problems that might occur in fitting linear regression lines are briefly discussed and references are given to more extensive information.

11.5.1 Use of Transformations in Linear Regression

Examination of the scatter diagram sometimes reveals that the points do not lie on a straight line. In Section 5.5, we mentioned taking transformations of the data in order to achieve a distribution that was approximately normally distributed. We discussed the use of the logarithmic and square root transformations if the observations were skewed to the right. These two transformations are often used to transform a curve into a straight line.

Figure 11.7 shows a hypothetical regression curve that is commonly found in biomedical examples. When curves such as these are found, it is possible to use either the logarithmic or perhaps a square-root transformation of X to achieve an approximate straight line. The logarithmic transformation will tend to have a greater effect than the square root. Both transformations have the effect of reducing large values more than small values, and hence tend to make either curve a straight line. In Figure 11.7, the arrows are drawn to show this reduction.

With statistical programs, it is a simple matter to try various transformations, and then obtain the scatter diagram of the transformed X and original Y. If a logarithmic

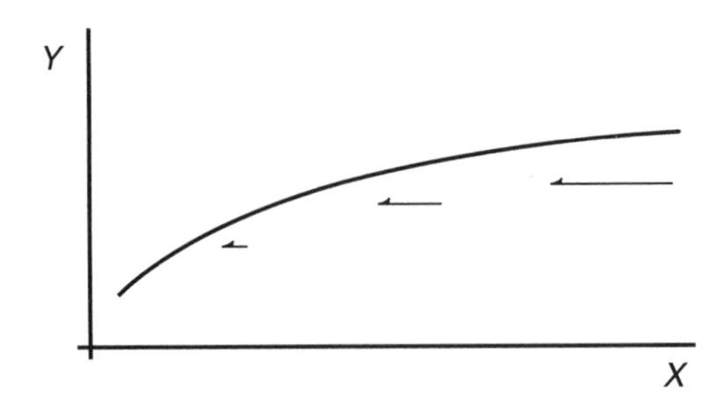

Fig. 11.7. Illustration of shape of a nonlinear regression curve.

transformation of X was successful, then the resulting regression equation would be

$$Y = a + b(\log_{10} X)$$

It is preferable to try a transformation on X rather than Y since the resulting equation is easier to interpret.

Additional information on the use of transformations in regression analysis can be found in Lewis-Beck [1980], Afifi and Clark [1996], Dunn and Clark [1987], or Atkinson [1985].

11.5.2 Effect of Outliers from the Regression Line

An outlier is an observation that deviates appreciably from the other observations in the sample. When a single variable X is measured, we often consider any observation that is a long way from the median or mean to be a possible outlier. Outliers can be caused by errors in measurement or in recording data or they may simply be unusual values. For example, a weight of 400 lbs may be a typing error or may be the weight of an unusually heavy person.

Concern about outliers in regression analysis arises because they may have a large effect on the estimate of a and b and consequently affect the fit of the line to the majority of the points. In linear regression, one of the best tools for finding outliers is to examine the scatter diagram. If one or two outliers result in a line not fitting the other points, then it is often advisable to check each outlier and consider removing it from the analysis.

Regression outliers have been classified as outliers in X, outliers in Y, and points that have a large effect on the slope of the line, often called *influential* points. Outliers in Y are located well above or well below the regression line. Many statistical programs print out the residuals from the line; outliers in Y may be detected by their large residuals.

Outliers in X are values that are far away from $\overline{X}$. Outliers in X possess the potential for having a large effect on the regression line. If a point is an outlier in X and is appreciably above or below the line, it is called an influential value since it can have a major effect on the slope and intercept. In general, a point that is an outlier in both X and Y tends to cause more problems in the fit of the line. For further discussion and illustrations, see Fox and Long [1990], Fox [1991], Chatterjee and Hadi [1988], or Afifi and Clark [1996].

Note that in general there are numerous excellent texts on regression analysis, so finding additional books to read is not a problem. Most of the books devoted to regression analysis also take up multiple regression analysis in which more than one X variable is used. This topic is not to be covered here.

PROBLEMS

11.1 The following rates are for all deaths from firearms per 100,000 persons in the United States for the years 1985–1995. The information was taken from the National Center for Health Statistics, *Health, United States 1996–97 and Injury Chartbook*.

Year	Death Rate from Firearms
1985	18.3
1986	19.7
1987	19.1
1988	20.5
1989	21.4
1990	23.6
1991	25.2
1992	24.9
1993	26.4
1994	26.0
1995	23.4

(a) State which model (single sample or fixed-X) this data set is.
(b) Plot a scatter diagram and compute the regression line.
(c) Compute 95% confidence intervals for β and test the H_0: $\beta = 0$.

11.2 Use the following data on length of service in days (X) and cost per day in dollars (Y) for 10 patients on a home health care program:

(a) State which model the data follow and plot the scatter diagram.
(b) Do you expect the r to be positive or negative from looking at the scatter diagram? Should b be positive or negative?
(c) Give a 95% confidence interval for ρ.
(d) Test the H_0: $\rho = 0$ and H_0: $\beta = 0$. Did you get the same P value?

Patient Number	Length of Service X	Cost per Day Y
1	310	35.00
2	89	34.60
3	22	102.60
4	9	136.20
5	120	69.50
6	99	79.60
7	63	140.20
8	170	45.40
9	20	81.3
10	198	29.50

11.3 In Problem 6.9 data are presented on counts of bacteria on 12 plates by two observers. Both observers counted the same plate and their counts were recorded. In Problem 6.9, a confidence interval was computed on the difference in the mean counts. The confidence limits did not include 0.

(a) Plot a scatter diagram of this data and compute r.
(b) Contrast what you can learn from a scatter plot and r with what you can learn from a confidence limit about the differences in the means.

11.4 The following table contains data from 17 countries. These countries are a systematic sample from twice this number of countries who had a population > 30 million people in 1998 and had complete data. The data were taken from the US Bureau of Census, *Statistical Abstracts of the United States; 1998*, 118th ed., Washington, DC. It consists of the crude birth rate (CBR) or the total number of births divided by the total population in 1998, life expectancy (LifeExp) from birth in years for the overall population, and gross domestic product (GDP) per capita for 1995. The GDP has been adjusted and then converted to United States dollars.

Country	CBR	LifeExp	GDP
Argentina	20.0	74.5	7,909
Brazil	20.9	64.4	4,080
Canada	12.1	79.2	19,000
Columbia	24.9	70.1	2,107
Egypt	27.3	62.1	746
France	11.7	78.5	26,290
India	25.9	62.9	348
Iran	31.4	68.3	2,449
Japan	10.3	80.0	41,160
Mexico	25.5	71.6	2,521
Pakistan	34.4	59.1	482
Poland	9.8	72.8	5,404
S. Africa	26.4	55.7	3,185
Tanzania	40.8	46.4	134
Thailand	16.8	69.0	2,806
United Kingdom	12.0	77.2	19,020
United States	14.4	76.1	27,550

(a) State which model the data in the table follows.
(b) Plot the scatter diagram when $X =$ life expectancy and $Y =$ crude birth rate and fit a regression line.
(c) Does the data appear to follow a straight line?
(d) If life expectancy increased by 10 years, what do you expect to happen to the crude birth rate?
(e) Test that H_0: $\beta = 0$. If you test that $\rho = 0$, will you get the same P value?

11.5 Plot a scatter diagram using GDP per capita as the X variable and life expectancy as the Y variable from the table in Problem 11.4.

(a) Fit a straight line to the plot of X = GDP and Y = LifeExp. Describe how the straight line fits the data.
(b) Compute r.
(c) Take the logarithm of X =GDP and repeat the steps in (a) and (b) using log(GDP) as the X variable.
(d) Explain why there is an increase in r when the logarithm of X is taken.

11.6 Plot the scatter diagram using the crude birth rate as the Y variable and GDP per capita as the X variable from the data given in Problem 11.4.

(a) Fit a least-squares regression line and compute r.
(b) Try a transformation on X and recompute r. Did it increase and if so why?

11.7 In Problem 11.4, a regression line was fitted using X = life expectancy and Y = crude birth rate. The $\overline{X}$ = 68.7 years and $\overline{Y}$ = 21.4 births per 1000 population. Here, the effect of the three types of outliers will be explored.

(a) Add an outlier in Y that has a value (X, Y) of (68.7,44) and recompute the regression line.
(b) Remove the outlier added in (a) and add an outlier in X of (85,11). Recompute the regression line.
(c) Remove the outlier in (b) and add an outlier in X and Y (influential point) of (85,44). Recompute the regression line.
(d) Compare the effects of these three outliers on the slope coefficient and the correlation coefficient.

REFERENCES

Afifi, A. A. and Clark, V. [1996]. *Computer-Aided Multivariate Analysis*, 3rd ed., London: Chapman & Hall, 109–111, 104–108.

Atkinson, A. C. [1985]. *Plots, Transformations and Regressions*, New York: University Oxford Press.

Chambers, J. M., Cleveland, W. S., Kleiner, B., and Tukey, P. A. [1983]. *Graphical Methods for Data Analysis*, Belmont, CA: Wadsworth Incorporated, 75–124.

Chatterjee, S. and Hadi, A. S. [1988]. *Sensitivity Analysis in Linear Regression*, New York: Wiley-Interscience, 71–182.

Cleveland, W. S. [1985]. *The Elements of Graphing Data*, Monterey, CA: Wadsworth Incorporated, 155–191.

Dunn, O. J. and Clark, V. A. [1987]. *Applied Statistics: Analysis of Variance and Regression*, 2nd ed. New York: Wiley, 280–281, 284–288, and 408–410.

Fisher, L. D. and van Belle, G. [1993]. *Biostatistics: A Methodology for the Health Sciences*, New York: Wiley-Interscience, 377–379.

Fox, J. and Long, J. S. [1990]. *Modern Methods of Data Analysis*, Newbury Park, CA: Sage, 257–291.

Fox, J. [1991]. *Regression Diagnostics*, Newbury Park, CA: Sage, 21–39.

Lewis-Beck, M. S. [1980]. *Applied Regression: An Introduction*, Newbury Park, CA: Sage.

CHAPTER TWELVE

Introduction to Survival Analysis

In biomedical applications, survival analysis is used to study the length of time until an *event* occurs. For example, survival analysis has been used in studying the length of time that cancer patients survive. Here, the event is death. For some diseases, such as multiple sclerosis, the length of time that the disease remains in remission has been analyzed. Survival analysis has also been used in studying the length of time women whose partners use condoms have remained nonpregnant.

What is called survival analysis in biomedical applications is called failure time or reliability analysis in engineering applications. In behavioral science, survival analysis has been used to analyze the length of time a person is on welfare or the time until a second arrest; it is called *event history* analysis. Here, we will use the terminology *survival analysis* whether or not the outcome being studied is death.

In Section 12.1, we discuss how the time to an event is measured and describe why survival data requires different statistical methods from those given in previous chapters. Section 12.2 presents graphical methods of depicting survival data. The death density, cumulative death distribution function, survival function, and hazard function are described and graphed. In Section 12.3, methods of estimating these functions are given using clinical life tables and the Kaplan–Meier method. Section 12.4 compares the use of the Kaplan–Meier method with clinical life tables and describes when each is useful. Section 12.5 briefly describes other analyzes performed using survival analysis.

12.1 SURVIVAL ANALYSIS DATA

12.1.1 Describing Time to an Event

In order to determine the time until an event, we need to define both a *starting* and an *ending time*. It is the starting time that usually presents more difficulties. Ideally, we want the starting time for all patients in the study to be the same in terms of their *course of disease*. If the starting time for patient A is early in the course of the disease due to a sensitive screening program and if, for patient B the starting time is after the patient has had the disease for a long time and after treatment is in progress, then obviously their starting times are not comparable. With data of this type, the results are difficult if not impossible to evaluate. Both in planning a study and in reporting the results, a clear statement of starting times is essential.

The *event* defining the end of the length of time is usually better known. Death certificates and hospital records provide clear statements of time of death. Precise time to an *event* such as remission may be more difficult to obtain for even the patient may be somewhat unsure when it has occurred. Usually, however, the event is clearly defined and its time accurately known.

For small samples, the time until an event occurs can be determined by counting the number of days or whatever unit of time is used from the starting time to the ending time for each patient by hand. But for large samples, especially when the time periods are long, computer programs are generally used. Statistical programs that compute both the length of time from starting to ending dates and also perform the analysis are ideal. When they are not available, some of the more complete spreadsheet programs will compute time periods between dates and can be used for some of the analyses.

12.1.2 Example of Measuring Time to an Event

We begin with a small hypothetical study to show how time to an event is measured. In this example, the event is death. In a typical survival study, the study starts at a particular point in time. Here, it will be assumed that the study starts enrolling patients at the beginning of 1995. Baseline data are collected on each patient at the time of enrollment. Then, treatment and follow-up starts and information is collected on the status of each patient. The study continues until terminated at the beginning of year 2003. Between 1995 and 2003, the starting time when each patient enters the study is considered to be when their disease condition is diagnosed. Note that researchers seldom have a large enough sample of patients immediately available and need to enter patients over time.

Patients also leave the study. This can occur because the patient died or was lost to follow-up or perhaps because the patient deliberately left the study. Some patients are not followed until death because the study ends in the year 2003.

Typical examples are shown for five patients in Figure 12.1 to illustrate these possibilities. Patient 1 started in 1995 and died 2 years later. Patient 2 entered in 1996 and died 3 years later. Patient 3 entered in 2001 and died 1 year later. Patient 4 entered in 1999 and was *lost to follow-up* after 2.5 years. At the time of the end of the study, it was not known what happened to this patient. Patient 5 entered in 1997 and was still alive at the end of the study. When this happens to a patient, the patient is said to be *withdrawn alive*.

The same data is shown in Figure 12.2 using the start of observation for each patient as the starting point. Here the results are shown as if all the patients entered at the same time. In pretending that all the patients entered at the same time, we assume that there were no changes over time either in the treatment or in the type of patient enrolled that would affect the time to death. Note that in Figure 12.2 it is easier to compare the times to an event than in Figure 12.2.

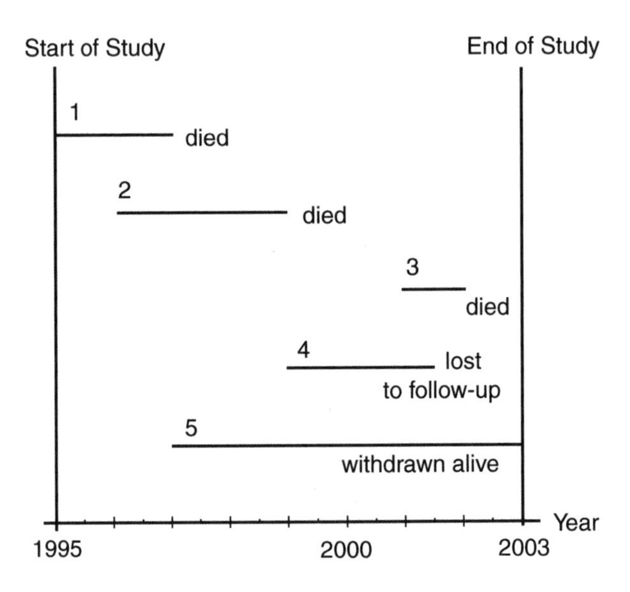

Fig. 12.1. Example of measuring time to an event for five patients.

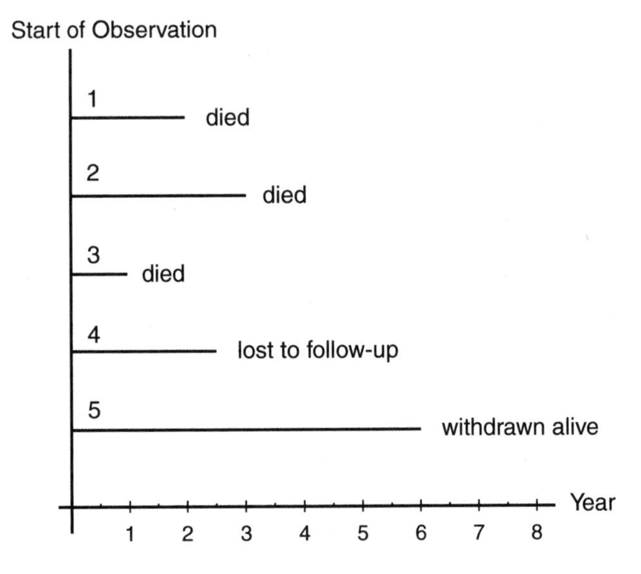

Fig. 12.2. Time to event with all patients starting at time 0.

In this example, the known dead are three. Further, we know that patient 4 lived at least 2.5 years and that patient 5 lived at least 6 years. These last two patients' observations are said to be *censored*. With censored data we know something about the survival time but we do not know the actual survival time.

One reason for examining the survival data from the five patients was to see why survival analysis is different from the methods already covered in this text for continuous (interval or ratio) data. If we have measurements of data such as heights, we do not have censored observations. Survival analysis enables us to use all the available data including the data on the patients who were lost to follow-up or withdrawn due to the study ending but whose actual time of death is unknown. If we use the methods given in previous chapters and simply ignore the lost or withdrawn patients, our results will be biased. These methods should not be used.

In Section 12.2, we present definitions and graphical descriptions of survival data.

12.2 SURVIVAL FUNCTIONS

In Section 3.2, we discussed displaying continuous data with histograms and distribution curves. We also gave instructions for computing cumulative frequencies and displaying the data given in Table 3.5. An example of a cumulative percent plot for the normal distribution was given in Figure 5.11. In this section, we will present graphical displays similar to those in Sections 3.2 and 5.1 and also present two somewhat different graphs commonly used in survival analysis. In order to simplify the presentation, we initially will ignore the topic of censored data.

12.2.1 The Death Density Function

We can imagine having data from a large number of patients who have died so that a histogram of the time to death could be made with very narrow class intervals. A frequency polygon could then be constructed similar to that given earlier in Figure 3.3. The vertical scale of the frequency polygon can be adjusted so the total area under the distribution curve is 1. Note that the total area under other distributions such as the normal distribution is also 1. When we are plotting time to an event such as death this curve is called the *death density function*. An example of a death density function is given in Figure 12.3. In Figure 12.3, the times until death range from 0 to approximately 3 years. The total area under the curve is 1. The death density function is helpful in assessing the peak time of death. Also, the shape of the death density function is often of interest. In most statistical books, the notation used to indicate the death density is $f(t)$ where t is used to denote time.

Note that the death density function in Figure 12.3 is skewed to the right. This is a common shape for such data. Survival data, unlike other continuous data that we have analyzed so far, are almost never assumed to be normally distributed.

In Figure 12.3, a vertical line has been drawn at a particular time t. For any given t, the area to the left of t represents the proportion of patients in the population who have died before time t. This proportion has been shaded gray. This area is called the *cumulative death distribution function* and is usually denoted by $F(t)$. The area to the right of t is $1 - F(t)$ since the total area under the curve in Figure 12.3 is 1. The area to the right represents the proportion of patients who have not died yet or in

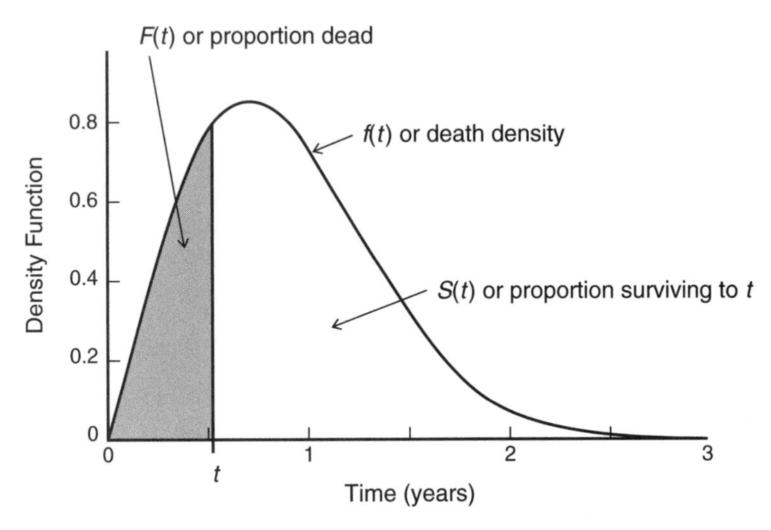

Fig. 12.3. Death density function, $f(t)$.

other words, the proportion *surviving* at least to time t. The proportion surviving at least to time t is denoted by $S(t)$.

12.2.2 The Cumulative Death Distribution Function

In Figure 12.4, the cumulative death distribution function $F(t)$ is displayed. Note that at time zero it is zero and it increase up to one at about time 3. The height of the curve at time t gives the proportion of patients who have died before time t. By time $t = 3$, all the patients have died. In Figure 12.4, t is approximately 0.8 and a little $< .5$ of the patients have died by this time. Note that $F(t)$ in Figure 12.4 looks rather similar to Figure 5.11 even though Figure 5.11 is the cumulative function for a symmetric normal curve. This illustrates the difficulty in distinguishing the shape of the death density function by simply looking at the cumulative death distribution function.

Figure 12.4 also illustrates the relationship between $F(t)$ and $S(t)$, the survival function. At any time t, the proportion dying before t plus the proportion surviving at least to t equals 1.

12.2.3 The Survival Function

Figure 12.5 displays the survival function. Note that this is simply Figure 12.4 flipped over. The survival function $S(t)$ starts at 1 at time 0 (all the patients are alive) and decreases to 0 at time 3. Sometimes the results are multiplied by 100 and reported in percent. We would say that 100% survived at time 0. Just as it was difficult to decide on the shape of the death density function by looking at the cumulative death distribution function, it is also difficult to use the survival function for this purpose.

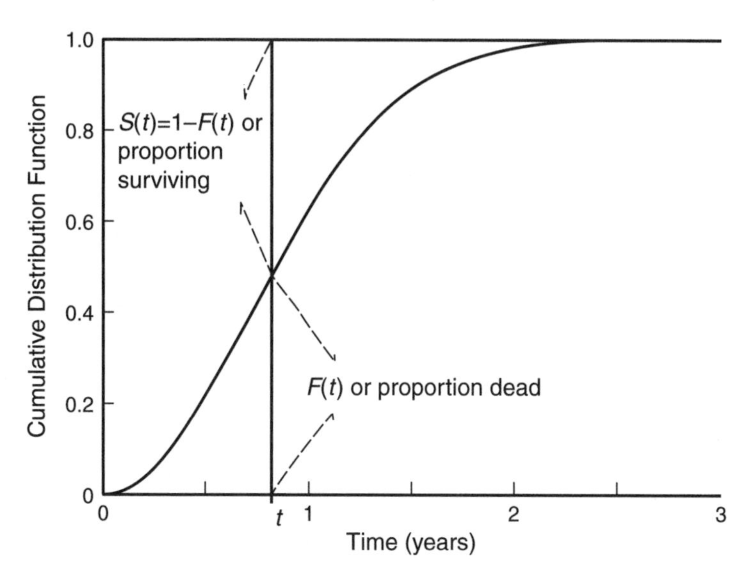

Fig. 12.4. Cumulative death density function, $F(t)$.

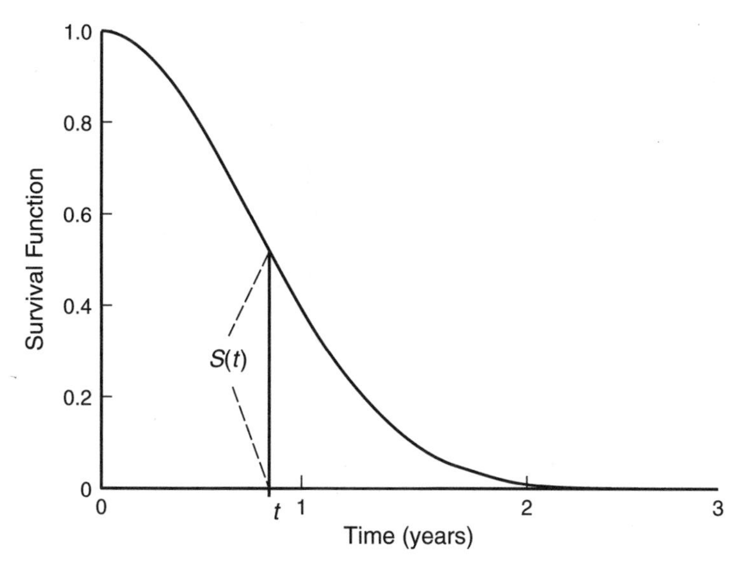

Fig. 12.5. Survival function, $S(t)$.

The survival function is useful in estimating the proportion of patients who will survive until time t. For example, patients with a life threatening disease might want to know what proportion of patients treated in a similar fashion live 1 year. From the survival function in Figure 12.5, we can see that about 40% survive at least 1 year. This can be seen by finding 1 year on the horizontal axis and drawing a vertical straight line that crosses the curve at a little less than 0.4 on the vertical axis. Or a physician might want to know how long the typical patient survives. This can be accomplished by finding the place on $S(t)$ where the height is 0.5, and then looking down to see the corresponding time.

If one treatment resulted in an appreciably higher survival function than another, then that would presumably be the preferred treatment. Also, the shape of the survival function is often examined. A heroic treatment that results in an appreciable number of patients' dying soon and thereafter followed by a very slowly declining survival function can be contrasted with a safer but ineffective treatment that does not result in many immediate deaths but whose survival rate keeps declining steeply over time.

Statistical survival programs commonly include a plot of the survival function in their output and it is used to evaluate the severity of illnesses and the efficacy of medical treatments.

12.2.4 The Hazard Function

One other function commonly used in survival analysis is the *hazard function*. In understanding the risk of death to patients over time, we want to be able to examine the risk of dying given that the patient has lived up to a particular time. With a severe treatment, there may be high risk of dying immediately after treatment. Or, as in some cancers, there may be a higher risk of dying two or more years after operation and chemotherapy.

The hazard function gives the conditional probability of dying between *time t and t plus a short interval called* Δt given survival at least to time t, all divided by Δt, as Δt approaches 0. The hazard function is not the chance or probability of a death but instead is a rate. The hazard function must be > 0 but there is no fixed upper value and it can be > 1. It is analogous to the concept of speed. Mathematically, the hazard function is equal to $f(t)/S(t)$ or the death density divided by the survival function. It is also called the force of mortality, conditional failure rate, or instantaneous death rate. We will denote the hazard function by $h(t)$.

One reason $h(t)$ is used is that its shape can differ markedly for different diseases. Whereas all survival functions, $S(t)$, are decreasing over time, the hazard function can take on a variety of shapes.

These shapes are used to describe the risk of death to patients over time. Figure 12.6 shows some typical hazard functions. The simplest shape labeled number 1 is a horizontal straight line. Here, the hazard function is assumed to be constant over time. A constant hazard function occurs when having survived up to any time t has no effect on the chance of dying in the next instant. Some authors call this the *no memory model*. It assumes that failures occur randomly over time. Although this assumption may appear unrealistic, a constant hazard rate has been used in some biomedical applications. For example, it has been shown that the length of time from the first to the second myocardial infarction can be approximated by a constant hazard function. When the hazard function is constant, a distribution called the exponential distribution can be assumed. The exponential is the simplest type of theoretical distribution used in survival analysis.

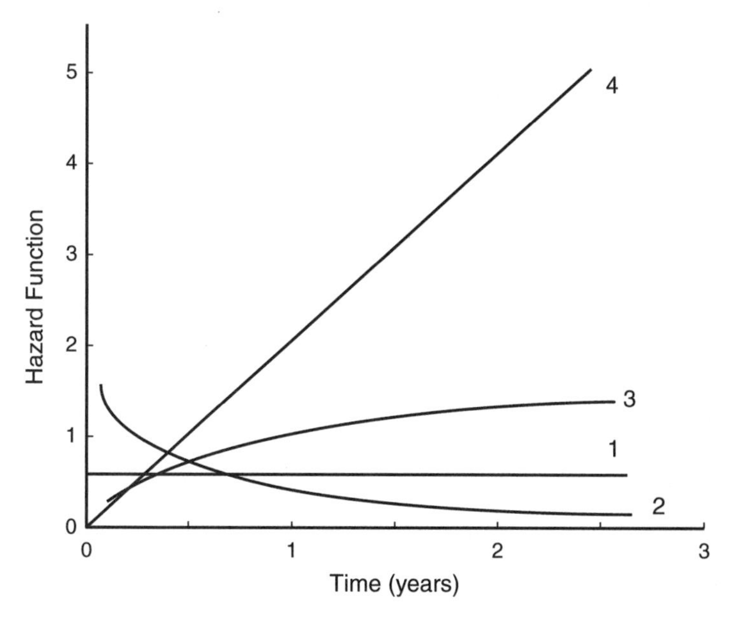

Fig. 12.6. Hazard functions, $h(t)$.

Numbers 2 and 3 in Figure 12.6 show a decreasing and an increasing hazard function. A decreasing hazard function is found when as time passes since the entry into the study the patient is more apt to live (at least for a short interval). This could occur because an operation is successful in treating the condition and the patients who die due to the operation do so soon after the operation. An increasing hazard function may be found when the treatment is not successful and as time passes the patient is more likely to die.

Number 4 shows a linear increasing hazard function. This is the hazard function from the death density and survival function given in Figures 12.3 and 12.5. That is, if we took a particular value of t, and divided the height of $f(t)$ in Figure 12.3 by the height of $S(t)$ at the same value of t, we would obtain a point on the linearly increasing line labeled Number 4. In interpreting $h(t)$ and $S(t)$, it is useful to keep in mind that they are inversely related. That is, if $S(t)$ quickly drops to a low value, then we would expect to have a high initial $h(t)$.

12.3 COMPUTING ESTIMATES OF $f(t)$, $S(t)$, AND $h(t)$

In this section, we will show how to estimate the death density, survival function, and hazard function using clinical life tables. In making life tables, the data are grouped in a manner somewhat similar to what was done in Section 3.1. We will also present estimation of the survival function using the Kaplan–Meier method. The

Kaplan–Meier method does not require the data to be grouped. These two descriptive methods are often used in order to obtain graphs of survival data.

12.3.1 Clinical Life Tables

We now demonstrate the method for computing clinical life tables using the data presented in Table 12.1. The data in Table 12.1 are a sample taken from the same distribution used to draw Figures 12.3–12.6; the distribution is called a Weibull distribution, a distribution widely used in survival analysis.

The sample size is 40. In the first column of Table 12.1 are listed the patient numbers. In the second column are listed the known days the patients lived after entering the study. In the third column, the last known status of the patient is given. This same pattern is repeated in the next two sets of three columns. The last known status has been coded 1 if the patient died, 2 if the patient was lost to follow-up, or 3 if the patient was withdrawn alive. For example, patient 1 died at 21 days and patient 5 was lost to follow-up at 141 days. Patients 20 and 37 were withdrawn alive.

In making a clinical life table, we must first decide on the number of intervals we want to display. With too many intervals, each one has very few individuals in it, and with too few intervals we lose information due to the coarse grouping.

Here we will choose intervals of 0.5 year, which will result in five intervals. At the start of the first interval, there are 40 patients entered in the study. The first interval goes from 1 to 182 days or up to 0.5 year and is written as 0.0 to $<$ 0.5 year. In computing clinical life tables no distinction is made between lost or withdrawn; their sum will be labeled c for censored. During the first interval eight patients die and one is censored (patient 5).

The second interval is 0.5 to $<$ 1.0 years or 183–365 days. During this time period, 15 patients died (2 were lost and 1 was withdrawn alive or 3 censored). The

Table 12.1. Patient Data for Clinical Life Table

Patient	Days	Status	Patient	Days	Status	Patient	Days	Status
1	21	1	15	256	2	29	398	1
2	39	1	16	260	1	30	414	1
3	77	1	17	261	1	31	420	1
4	133	1	18	266	1	32	468	2
5	141	2	19	269	1	33	483	1
6	152	1	20	287	3	34	489	1
7	153	1	21	295	1	35	505	1
8	161	1	22	308	1	36	539	1
9	179	1	23	311	1	37	565	3
10	184	1	24	321	2	38	618	1
11	197	1	25	326	1	39	793	1
12	199	1	26	355	1	40	794	1
13	214	1	27	361	1			
14	228	1	28	374	1			

Table 12.2. Computations for a Clinical Life Table

Interval	n_{ent}	c	n_{exp}	d	$\hat{q}$	$\hat{p}$	$\hat{S}(t)$	$\hat{f}(t)$	$\hat{h}(t)$
0.0 to < 0.5	40	1	39.5	8	.203	.797	1.000	.406	.451
0.5 to < 1.0	31	3	29.5	15	.508	.492	.797	.810	1.364
1.0 to < 1.5	13	1	12.5	8	.640	.360	.392	.502	1.882
1.5 to < 2.0	4	1	3.5	1	.286	.714	.141	.081	.667
2.0 to < 2.5	2	0	2	2	1.000	.000	.101	.202	4.000

next interval is 1.0 to < 1.5 years and has 8 patients who die and 1 was censored. The next interval, 1.5 to 2.0 has 1 who died and 1 censored, and the last interval has 2 patients who died.

The results are summarized in Table 12.2. The first column describes the intervals; note that the width of the interval is $w = 0.5$ years. The second column, labeled n_{ent}, gives the number of patients entering each interval. The third column labeled c includes a count of the patients who are censored. The fifth column lists the number of patients who die in each interval and is labeled d. This set of columns displays the information given in Table 12.1.

Before discussing the remaining columns of Table 12.2, we show how to compute the number of patients entering each interval (n_{ent}). For the first interval, it is the sample size, here 40. For the second interval, we begin with the n_{ent} for the first interval and subtract from it both the number dying and the number censored during the first interval. That is, we take $40 - 1 - 8 = 31$ starting the second interval. For the start of the third interval, we take $31 - 3 - 15 = 13$. The numbers entering the fourth and fifth interval are computed in the same fashion. In general, from the second interval on, the number at the outset of the interval is the number available at the outset of the previous interval minus the number who are censored or die during the previous interval.

The remaining columns in the clinical life table are obtained by performing calculations on the columns previously filled in. The column labeled n_{exp} gives the number exposed to risk. It is computed as

$$n_{\text{exp}} = n_{\text{ent}} - c/2$$

If there are no censored patients, the number exposed to risk is the number entering the interval. If there is censoring, it is assumed that the censoring occurs evenly distributed throughout the interval. Thus on the average, the censored patients are assumed to be censored from the study for one-half of the total interval. Hence, the number censored in each interval is divided by 2 in estimating the number exposed to risk. The number exposed to risk decreases in a clinical life table with successive intervals as long as at least one patient dies or is censored. For the first interval, $n_{\text{exp}} = 40 - 1/2 = 39.5$ since one patient is censored. For the second interval, $n_{\text{exp}} = 31 - 3/2 = 29.5$, and so on. Note that in the last row or two of Table 12.2, there are very few patients left in the sample. The estimates obtained from the final row may be quite variable.

In survival analysis, the number of patients exposed to risk decreases either as patients die or are censored. In the relative frequency Table 3.5, all the percents or proportions are computed using the original sample size n. Here, the following computations will use the number actually exposed to risk not simply the number entering the study.

The next column is labeled $\hat{q}$. The hat over the q is there to indicate that $\hat{q}$ is an approximation. It estimates the proportion of patients who die in an interval given that they are exposed to risk in that interval. It is computed from

$$\hat{q} = d/n_{\text{exp}}$$

For example, for the first interval we compute $8/39.5 = .203$. For the second interval, we have $15/29.5 = .508$.

The column labeled $\hat{p}$ is computed simply as

$$\hat{p} = 1 - \hat{q}$$

and is the proportion of patients who survive an interval given that they are exposed to risk in the interval. For the first interval, $\hat{p} = 1 - .203 = .797$.

The survival function gives the proportion surviving up to the *start* of the interval. In some texts, it is denoted by $\hat{P}(t)$ rather than the $\hat{S}(t)$ used in this book. The sample survival function, $\hat{S}(t)$, for the first interval is equal to 1 since all patients survive up to the beginning of the first interval. For the remaining intervals, it is computed by multiplying $\hat{p}$ by $\hat{S}(t)$, both from the preceding interval. For example, for $\hat{S}(t)$ for the second interval we multiply .797 by 1.000 to obtain $\hat{S}(t) = .797$ for the second interval. For the third interval, we multiply .492 by .797 to obtain .392. For the fourth interval, we compute $.360(.392) = .141$, and the last interval is $.714(.141) = .101$. In other words, the chance of surviving to the start of a particular interval is equal to the chance of surviving up to the start of the preceding interval times the chance of surviving though the preceding interval. In graphing $\hat{S}(t)$ on the vertical axis, the computed values are graphed above the start of the interval on the horizontal axis.

The sample death density, $\hat{f}(t)$, is estimated at the *midpoint* of each interval. For each interval, we compute

$$\hat{f}(t) = \frac{\hat{q}\hat{S}(t)}{w}$$

where w is the width of the interval. In this example, $w = .5$. For example, for the first interval, we compute $1(.203)/.5 = .406$. For the second interval, we have $.797(.508)/.5 = .810$. The other intervals proceed in a similar fashion.

Finally, the estimate of the sample hazard function, $\hat{h}(t)$, is also plotted at the *midpoint* of each interval. The formula for this estimate is

$$\hat{h}(t) = \frac{d}{w[n_{\text{exp}} - d/2]}$$

Since the hazard is computed at the midpoint of the interval, we subtract one-half of the deaths from n_{exp}. Here, we are assuming that the deaths occur in a uniform fashion throughout the interval, so at the midpoint of the interval one-half of them will have occurred. We multiply the denominator by the width of the interval to get the proper rate. For example, for the first interval we have

$$\hat{h}(t) = \frac{8}{.5[39.5 - 8/2]} = \frac{8}{.5[35.5]} = .451$$

The numerical estimate of the hazard function for the second interval would be $15/(.5[29.5 - 15/2]) = 1.364$. The remaining intervals are computed in the same fashion.

If a statistical program is used to compute the life table, it will likely contain other output. This might include standard errors of the estimates of the death density, survival function, and hazard function (see Fisher and van Belle [1993] or Gross and Clark [1975]). Also, some programs will compute the 50% survival time from the column labeled $\hat{S}(t)$ by interpolation. For example, in Table 12.2 we know that at the beginning of the second interval .797 have survived, and at the beginning of the third interval .392 have survived. Thus, at some time between .5 and 1.0 years, 0.50 or 50% of the patients survived. This can be computed using the linear interpolation formula given in Section 5.2.2.

12.3.2 Kaplan–Meier Estimate

In Section 12.3.1 the computations for the clinical life table were presented. Note that in making these tables we grouped the data into one-half year intervals even though we knew the number of days each patient lived. The *Kaplan–Meier* or *product limit* method of estimating the survival function uses the actual length of time to the outcome event such as death or to censoring due to loss to follow-up or withdrawn alive from the study.

As might be expected, the Kaplan–Meier method is considerably more work to compute by hand if the sample size is large and so statistical programs are usually used to perform the computations and graph the results. Here, we will illustrate the computations with a small example.

A study was made of 8 patients who were admitted to a hospital with a life threatening condition. The outcome event is death. Two patients were transferred from the hospital before they died and they are considered to be censored at hospital discharge.

Table 12.3. Computations for Kaplan–Meier Method

Days	Deaths	Censored	n_{obs}	$(n_{\text{obs}} - d)/n_{\text{obs}}$	$\hat{S}(t)$
2	1		8	$(8 - 1)/8 = .875$	.875
3	1		7	$(7 - 1)/7 = .857$	.750
3		1	6		
4	2		5	$(5 - 2)/5 = .600$	.450
5		1	3		
7	1		2	$(2 - 1)/2 = .500$	.225
9	1		1	$(1 - 1)/1 = .000$	0

The first step in analyzing the data is to order the observations from smallest to largest. The ordered times were 2, 3, 3^c, 4, 4, 5^c, 7, and 9 days. The c indicates that the third and sixth patient were censored.

If the event and censoring occur at the same time as happened with day 3, Kaplan and Meier recommend treating the event as if it occurred slightly before the censoring and the censoring is treated as if it occurred slightly after the particular event.

The computations are summarized in Table 12.3. The column labeled *days* signifies the day the event occurred, the column labeled *death* gives the number of patients who died on a particular day, *censored* signifies lost to follow-up or withdrawn alive, n_{obs} is the number of patients observed on that day, $(n_{\text{obs}} - d)/n_{\text{obs}}$ is a computed quantity, and finally $\hat{S}(t)$ is the estimated survival function. In our example, $\hat{S}(t) = 1$ up to day 2 when the first patient died since all eight patients survived to day 2.

On day 2, one patient died and so the chance of surviving past 2 days is 7/8 or .875. The $(n_{\text{obs}} - d)/n_{\text{obs}}$ gives the proportion surviving at each particular time that a patient dies given they have survived up to then. The chance of surviving up to day 2 is 1 and the chance of surviving past day 2 is .875 given survival up to day 2, so the chance of surviving past day 2 is $\hat{S}(2) = .875(1) = .875$. Here, 2 has been substituted for t in $\hat{S}(t)$ since we are giving the results for day 2.

On day 3, one patient dies and one is censored. Note that we assume that the death occurs first. Thus, the patient dies from the seven remaining patients, so the chance of surviving on day 3 given the patient survived up to day 3 is $(7 - 1)/7 = .857$. Thus, the chance of surviving past day 3 is .857 times the chance of surviving up to day 3 or the chance of surviving past day 3 is $.857(.875) = .750$. There is no change in $\hat{S}(t)$ when a patient is censored and so no calculations are required in the third row except to reduce the number of patients observed by one to account for the patient who is censored. Thus at the start of day 4, we have only five patients.

On day 4, two patients die, and so the chance of dying on day 4 given the patient is known to be alive up to day 4 is $(5 - 2)/5 = .600$. The chance of surviving past day 4 is $\hat{S}(4) = .600(.750) = .450$, where .750 is the chance of surviving past day 3.

The remaining rows are computed in a similar fashion. The general formula for any given row can be given as $(n_{\text{obs}} - d)/n_{\text{obs}}$ times the numerical value of $\hat{S}(t)$ for the preceding row.

If we were to plot this data as a step function, we would first make a vertical axis that goes from 0 to 1 and a horizontal axis that goes from 0 to 10. The values of $\hat{S}(t)$ are plotted on the vertical axis and time in days is plotted on the horizontal axis. Between day 0 and day 2, we would plot a horizontal line that had a height of 1. From day 2 to day 3, another horizontal line would be plotted with a height of .875, from day 3 to 4, another horizontal line would have height of .750, and from day 4 to 7, a horizontal line with a height of .450 is plotted. On day 7 to day 9 the height of the horizontal line would be .225, and after day 9 the height would be 0.

For actual data sets with more times to the event, statistical programs are recommended both for the computations and for graphing the survival function as the Kaplan–Meier product limit method is considerable work to calculate and graph by hand.

12.4 COMPARISON OF CLINICAL LIFE TABLES AND THE KAPLAN–MEIER METHOD

The Kaplan–Meier method is recommended when the sample size is small. It also has the advantage that the user does not have to decide on the length of the interval as must be done for clinical life tables. Many statistical programs that include survival analysis in their output will display an estimate of the survival function using the Kaplan–Meier method. It is also possible to compute a mean survival time and the variance of the survival time using methods provided by Kaplan and Meier (see Gross and Clark [1975]) or a median (see Miller [1981]). Note that the mean survival time is sometimes difficult to interpret because survival distributions can be highly skewed. Some statistical programs will also print out the standard error of the estimated survival function at each distinct time of death.

The clinical life table method is useful for larger sample sizes. It also has the advantage of directly furnishing estimates of the hazard function and the death density function. It is possible to compute the standard errors of the estimated survival function, hazard function, and death density function at each interval.

Statistical programs often give the median survival time or it can be easily estimated from the clinical life table. If the investigators lack access to a survival program that produces a clinical life table, some spreadsheet programs can be used to perform much of the work. The time in say, days, between dates may be obtained from spreadsheet programs for each patient. The times can be sorted so it is straightforward to obtain the counts. Actually, it is simply necessary to know only which interval each length of time falls into. The last five columns of Table 12.2 are arithmetic manipulations of previous columns. Spreadsheet programs often provide the capabilities to perform the calculations to obtain these columns from the previous ones.

When the sample size is large, the intervals can be made quite small and the difference in appearance of a plot of the estimated survival function from a clinical life table or from the Kaplan–Meier method is negligible.

12.5 ADDITIONAL ANALYSES USING SURVIVAL DATA

In this section, we briefly mention two other types of analyses that can be performed using survival data that include censored observations. References are given to texts that provide additional information.

12.5.1 Comparing the Equality of Survival Functions

In addition to examining survival for a single group of patients, often we wish to compare survival for two or more treatment groups. Visually examining the estimate survival function or hazard rate is often done in order to compare a standard to an experimental treatment. If one survival curve is appreciably above the other, then usually the treatment with the higher curve is preferred. One difficulty in looking for differences in estimated survival functions is that minor differences in estimated survival functions sometime look larger than they are. For example, if there is a minor difference soon after entry to the study and if after that the survival functions are similar, the treatment that did better initially will tend to have a higher survival function throughout. This may exaggerate minor differences. It is useful to examine both the survival function and the hazard function to get a better idea of what happens over time.

Survival statistical programs may also offer the option of several tests for the null hypothesis that the two or more survival functions are equal. These tests were not derived to test for equality of the means as the tests described in Chapter 7 were. If the null hypothesis is rejected we can conclude that the survival functions are not equal.

For any of these tests, the test results should be treated with caution if the sample size is small. Also, the pattern of censoring should be examined to look for major differences between the treatment groups. For example, if the experimental group had a much higher rate of patients refusing treatment or being lost to follow-up, it may not be better than the standard treatment even though the few patients that remain on the experimental treatment do somewhat better than the larger proportion remaining on the standard treatment. It is always recommended that a plot be examined of time versus the number or proportion censored for both groups to see if there are major differences.

Two of the more commonly used tests are the *log-rank* test (also called the Mantel–Cox test) and the *Peto* test (also called the Peto–Breslow test). Statistical programs are commonly used to perform the tests and the description of the calculations is beyond the scope of this book (see Kleinbaum [1996] for a very understandable explanation). In considering these tests it should be kept in mind that the log-rank test places more emphasis on the results in the right tail of the survival function where the number at risk may be small and the Peto test places more emphasis at the beginning of the survival curve.

12.5.2 Regression Analysis of Survival Data

Regression analysis is also performed with survival data. The survival time is used as an *outcome* variable that can be predicted by a predictor variable or variables. This has become one of the major forms of survival analysis and is available in many statistical programs. For further information on this topic see Kleinbaum [1996], Afifi and Clark [1996], Fisher and van Belle [1993], or Allison [1984]. For other general texts on survival analysis that cover both this topic plus survival analysis in general, see Lee [1992] or Parmar and David [1995].

PROBLEMS

12.1 Plot the estimated survival function from the information given in Table 12.3.

12.2 Plot the estimated survival function, hazard function, and death density function from Table 12.2. Compare the results to those in Figures 12.5, 12.6 (number 4), and 12.3, respectively. Does it seem reasonable that the information given in Table 12.2 could be a sample from the population depicted in these figures?

12.3 Compute the median survival time for the data given in Table 12.2 using the linear interpolation formula given in Section 5.2.2.

12.4 If we ignored the censored observations and the computed mean survival time using the usual formula for the mean and only the observations from those who died, would our estimate of the true mean be too large or too small?

12.5 The following data are survival times in days from a life threatening condition: 15, 18, 18, 21, 21^c, 25^c, 26. Compute the estimated $\hat{S}(t)$ using the Kaplan–Meier method.

12.6 Data from a study of the pregnancy rate from couples using latex condoms is given in the following table. The data were obtained from the California Family Health Council, Inc. and used with their permission. The time since study entry has been grouped by month. The first two columns of the table gives the six time periods in months. The couples are followed for 6 months since this time period has been found to be about the maximum that couples are willing to fill out the daily diaries that are used to obtain information needed in other parts of the study. Instead of death, the outcome event is pregnancy. Compute the estimates of $S(t)$ and $h(t)$.

Pregnancy Data for Couples using Latex Condoms (months)

Equal to or	Less Than	Entered	Withdrawn	Lost	Pregnant
0–	1.0	384	23	2	4
1.0–	2.0	355	12	3	3
2.0–	3.0	337	16	4	4
3.0–	4.0	313	14	1	3
4.0–	5.0	295	5	1	2
5.0–	6.0	287	6	2	4

REFERENCES

Afifi, A. A. and Clark, V. [1996]. *Computer-Aided Multivariate Analysis,* 3rd Ed., London: Chapman & Hall, 317–327.

Allison, P. D. [1984]. *Event History Analysis: Regression for Longitudinal Event Data,* Newbury Park, CA: Sage, 17–66.

Fisher, L. D. and van Belle, G. [1993]. *Biostatistics: A Methodology for the Health Sciences,* New York: Wiley-Interscience, 786–827.

Gross, A. J. and Clark, V. A. [1975]. *Survival Distributions: Reliability Applications in the Biomedical Sciences,* New York: Wiley, 34–44.

Kleinbaum, D. G. [1996]. *Survival Analysis: A Self-Learning Text,* New York: Springer, 4–120.

Lee, E. T. [1992]. *Statistical Methods for Survival Data Analysis,* 2nd ed., New York: Wiley.

Miller, R. G. 1981. *Survival Analysis,* New York: Wiley, 74–76.

Parmar, M. K. B. and David, M. [1995]. *Survival Analysis: A Practical Approach*, New York: Wiley.

APPENDIX A

Statistical Tables

Table A.1. Random Numbers

10 09 73 25 33	76 52 01 35 86	34 67 35 48 76	80 95 90 91 17	39 29 27 49 45
37 54 20 48 05	64 89 47 42 96	24 80 52 40 37	20 63 61 04 02	00 82 29 16 65
08 42 26 89 53	19 64 50 93 03	23 20 90 25 60	15 95 33 47 64	35 08 03 36 06
99 01 90 25 29	09 37 67 07 15	38 31 13 11 65	88 67 67 43 97	04 43 62 76 59
12 80 79 99 70	80 15 73 61 47	64 03 23 66 53	98 95 11 68 77	12 17 17 68 33
66 06 57 47 17	34 07 27 68 50	36 69 73 61 70	65 81 33 98 85	11 19 92 91 70
31 06 01 08 05	45 57 18 24 06	35 30 34 26 14	86 79 90 74 39	23 40 30 97 32
85 26 97 76 02	02 05 16 56 92	68 66 57 48 18	73 05 38 52 47	18 62 38 85 79
63 57 33 21 35	05 32 54 70 48	90 55 35 75 48	28 46 82 87 09	83 49 12 56 24
73 79 64 57 53	03 52 96 47 78	35 80 83 42 82	60 93 52 03 44	35 27 38 84 35
98 52 01 77 67	14 90 56 86 07	22 10 94 05 58	60 97 09 34 33	50 50 07 39 98
11 80 50 54 31	39 80 82 77 32	50 72 56 82 48	29 40 52 42 01	52 77 56 78 51
83 45 29 96 34	06 28 89 80 83	13 74 67 00 78	18 47 54 06 10	68 71 17 78 17
88 68 54 02 00	86 50 75 84 01	36 76 66 79 51	90 36 47 64 93	29 60 91 10 62
99 59 46 73 48	87 51 76 49 69	91 82 60 89 28	93 78 56 13 68	23 47 83 41 13
65 48 11 76 74	17 46 85 09 50	58 04 77 69 74	73 03 95 71 86	40 21 81 65 44
80 12 43 56 35	17 72 70 80 15	45 31 82 23 74	21 11 57 82 53	14 38 55 37 63
74 35 09 98 17	77 40 27 72 14	43 23 60 02 10	45 52 16 42 37	96 28 60 26 55
69 91 62 68 03	66 25 22 91 48	36 93 68 72 03	76 62 11 39 90	94 40 05 64 18
09 89 32 05 05	14 22 56 85 14	46 42 75 67 88	96 29 77 88 22	54 38 21 45 98
91 49 91 45 23	68 47 92 76 86	46 16 28 35 54	94 75 08 99 23	37 08 92 00 48
80 33 69 45 98	26 94 03 68 58	70 29 73 41 35	53 14 03 33 40	42 05 08 23 41
44 10 48 19 49	85 15 74 79 54	32 97 92 65 75	57 60 04 08 81	22 22 20 64 13
12 55 07 37 42	11 10 00 20 40	12 86 07 46 97	96 64 48 94 39	28 70 72 58 15
63 60 64 93 29	16 50 53 44 84	40 21 95 25 63	43 65 17 70 82	07 20 73 17 90
61 19 69 04 46	26 45 74 77 74	51 92 43 37 29	65 39 45 95 93	42 58 26 05 27
15 47 44 52 66	95 27 07 99 53	59 36 78 38 48	82 39 61 01 18	33 21 15 94 66
94 55 72 85 73	67 89 75 43 87	54 62 24 44 31	91 19 04 25 92	92 92 74 59 73
42 48 11 62 13	97 34 40 87 21	16 86 84 87 67	03 07 11 20 59	25 70 14 66 70
23 52 37 83 17	73 20 88 98 37	68 93 59 14 16	26 25 22 96 63	05 52 28 25 62
04 49 35 24 94	75 24 63 38 24	45 86 25 10 25	61 96 27 93 35	65 33 71 24 72
00 54 99 76 54	64 05 18 81 59	96 11 96 38 96	54 69 28 23 91	23 28 72 95 29
35 96 31 53 07	26 89 80 93 54	33 35 13 54 62	77 97 45 00 24	90 10 33 93 33
59 80 80 83 91	45 42 72 68 42	83 60 94 97 00	13 02 12 48 92	78 56 52 01 06
46 05 88 52 36	01 39 09 22 86	77 28 14 40 77	93 91 08 36 47	70 61 74 29 41
32 17 90 05 97	87 37 92 52 41	05 56 70 70 07	86 74 31 71 57	85 39 41 18 38
69 23 46 14 06	20 11 74 52 04	15 95 66 00 00	18 74 39 24 23	97 11 89 63 38
19 56 54 14 30	01 75 87 53 79	40 41 92 15 85	66 67 43 68 06	84 96 28 52 07
45 15 51 49 38	19 47 60 72 46	43 66 79 45 43	59 04 79 00 33	20 82 66 95 41
94 86 43 19 94	36 16 81 08 51	34 88 88 15 53	01 54 03 54 56	05 01 45 11 76
98 08 62 48 26	45 24 02 84 04	44 99 90 88 96	39 09 47 34 07	35 44 13 18 80
33 18 51 62 32	41 94 15 09 49	89 43 54 85 81	88 69 54 19 94	37 54 87 30 43
80 95 10 04 06	96 38 27 07 74	20 15 12 33 87	25 01 62 52 98	94 62 46 11 71
79 75 24 91 40	71 96 12 82 96	69 86 10 25 91	74 85 22 05 39	00 38 75 95 79
18 63 33 25 37	98 14 50 65 71	31 01 02 46 74	05 45 56 14 27	77 93 89 19 36
74 02 94 39 02	77 55 73 22 70	97 79 01 71 19	52 52 75 80 21	80 81 45 17 48
54 17 84 56 11	80 99 33 71 43	05 33 51 29 69	56 12 71 92 55	36 04 09 03 24
11 66 44 98 83	52 07 98 48 27	59 38 17 15 39	09 97 33 34 40	88 46 12 33 56
48 32 47 79 28	31 24 96 47 10	02 29 53 68 70	32 30 75 75 46	15 02 00 99 94
69 07 49 41 38	87 63 79 19 76	35 58 40 44 01	10 51 82 16 15	01 84 87 69 38

Table A.1. (*Continued*)

09 18 82 00 97	32 82 53 95 27	04 22 08 63 04	83 38 98 73 74	64 27 85 80 44
90 04 58 54 97	51 98 15 06 54	94 93 88 19 97	91 87 07 61 50	68 47 66 46 59
73 18 95 02 07	47 67 72 62 69	62 29 06 44 64	27 12 46 70 18	41 36 18 27 60
75 76 87 64 90	20 97 18 17 49	90 42 91 22 72	95 37 50 58 71	93 82 34 31 78
54 01 64 40 56	66 28 13 10 03	00 68 22 73 98	20 71 45 32 95	07 70 61 78 13
08 35 86 99 10	78 54 24 27 85	13 66 15 88 73	04 61 89 75 53	31 22 30 84 20
28 30 60 32 64	81 33 31 05 91	40 51 00 78 93	32 60 46 04 75	94 11 90 18 40
53 84 08 62 33	81 59 41 36 28	51 21 59 02 90	28 46 66 87 95	77 76 22 07 91
91 75 75 37 41	61 61 36 22 69	50 26 39 02 12	55 78 17 65 14	83 48 34 70 55
89 41 59 26 94	00 39 75 83 91	12 60 71 76 46	48 94 97 23 06	94 54 13 74 08
77 51 30 38 20	86 83 42 99 01	68 41 48 27 74	51 90 81 39 80	72 89 35 55 07
19 50 23 71 74	69 97 92 02 88	55 21 02 97 73	74 28 77 52 51	65 34 46 74 15
21 81 85 93 13	93 27 88 17 57	05 68 67 31 56	07 08 28 50 46	31 85 33 84 52
51 47 46 64 99	68 10 72 36 21	94 04 99 13 45	42 83 60 91 91	08 00 74 54 49
99 55 96 83 31	62 53 52 41 70	69 77 71 28 30	74 81 97 81 42	43 86 07 28 34
33 71 34 80 07	93 58 47 28 69	51 92 66 47 21	58 30 32 98 22	93 17 49 39 72
85 27 48 68 93	11 30 32 92 70	28 83 43 41 37	73 51 59 04 00	71 14 84 36 43
84 13 38 96 40	44 03 55 21 66	73 85 27 00 91	61 22 26 05 61	62 32 71 84 23
56 73 21 62 34	17 39 59 61 31	10 12 39 16 22	85 49 65 75 60	81 60 41 88 80
65 13 85 68 06	87 64 88 52 61	34 31 36 58 61	45 87 52 10 69	85 64 44 72 77
38 00 10 21 76	81 71 91 17 11	71 60 29 29 37	74 21 96 40 49	65 58 44 96 98
37 40 29 63 97	01 30 47 75 86	56 27 11 00 86	47 32 46 26 05	40 03 03 74 38
97 12 54 03 48	87 08 33 14 17	21 81 53 92 50	75 23 76 20 47	15 50 12 95 78
21 82 64 11 34	47 14 33 40 72	64 63 88 59 02	49 13 90 64 41	03 85 65 45 52
73 13 54 27 42	95 71 90 90 35	85 79 47 42 96	08 78 98 81 56	64 69 11 92 02
07 63 87 79 29	03 06 11 80 72	96 20 74 41 56	23 82 19 95 38	04 71 36 69 94
60 52 88 34 41	07 95 41 98 14	59 17 52 06 95	05 53 35 21 39	61 21 20 64 55
83 59 63 56 55	06 95 89 29 83	05 12 80 97 19	77 43 35 37 83	92 30 15 04 98
10 85 06 27 46	99 59 91 05 07	13 49 90 63 19	53 07 57 18 39	06 41 01 93 62
39 82 09 89 52	43 62 26 31 47	64 42 18 08 14	43 80 00 93 51	31 02 47 31 67
59 58 00 64 78	75 56 97 88 00	88 83 55 44 86	23 76 80 61 56	04 11 10 84 08
38 50 80 73 41	23 79 34 87 63	90 82 29 70 22	17 71 90 42 07	95 95 44 99 53
30 69 27 06 68	94 68 81 61 27	56 19 68 00 91	82 06 76 34 00	05 46 26 92 00
65 44 39 56 59	18 28 82 74 37	49 63 22 40 41	08 33 76 56 76	96 29 99 08 36
27 26 75 02 64	13 19 27 22 94	07 47 74 46 06	17 98 54 89 11	97 34 13 03 58
91 30 70 69 91	19 07 22 42 10	36 69 95 37 28	28 82 53 57 93	28 97 66 62 52
68 43 49 46 88	84 47 31 36 22	62 12 69 84 08	12 84 38 25 90	09 81 59 31 46
48 90 81 58 77	54 74 52 45 91	35 70 00 47 54	83 82 45 26 92	54 13 05 51 60
06 91 34 51 97	42 67 27 86 01	11 88 30 95 28	63 01 19 89 01	14 97 44 03 44
10 45 51 60 19	14 21 03 37 12	91 34 23 78 21	88 32 58 08 51	43 66 77 08 83
12 88 39 73 43	65 02 76 11 84	04 28 50 13 92	17 97 41 50 77	90 71 22 67 69
21 77 83 09 76	38 80 73 69 61	31 64 94 20 96	63 28 10 20 23	08 81 64 74 49
19 52 35 95 15	65 12 25 96 59	86 28 36 82 58	69 57 21 37 98	16 43 59 15 29
67 24 55 26 70	35 58 31 65 63	79 24 68 66 86	76 46 33 42 22	26 65 59 08 02
60 58 44 73 77	07 50 03 79 92	45 13 42 65 29	26 76 08 36 37	41 32 64 43 44
53 85 34 13 77	36 06 69 48 50	58 83 87 38 59	49 36 47 33 31	96 24 04 36 42
24 63 73 87 36	74 38 48 93 42	52 62 30 79 92	12 36 91 86 01	03 74 28 38 73
83 08 01 24 51	38 99 22 28 15	07 75 95 17 77	97 37 72 75 85	51 97 23 78 67
16 44 42 43 34	36 15 19 90 73	27 49 37 09 39	85 13 03 25 52	54 84 65 47 59
60 79 01 81 57	57 17 86 57 62	11 16 17 85 76	45 81 95 29 79	65 13 00 48 60

Table A.1. ***(Continued)***

03 99 11 04 61	93 71 61 68 94	66 08 32 46 53	84 60 95 82 32	88 61 81 91 61
38 55 59 55 54	32 88 65 97 80	08 35 56 08 60	29 73 54 77 62	71 29 92 38 53
17 54 67 37 04	92 05 24 62 15	55 12 12 92 81	59 07 60 79 36	27 95 45 89 09
32 64 35 28 61	95 81 90 68 31	00 91 19 89 36	76 35 59 37 79	80 86 30 05 14
69 57 26 87 77	39 51 03 59 05	14 06 04 06 19	29 54 96 96 16	33 56 46 07 80
24 12 26 65 91	27 69 90 64 94	14 84 54 66 72	61 95 87 71 00	90 89 97 57 54
61 19 63 02 31	92 96 26 17 73	41 83 95 53 82	17 26 77 09 43	78 03 87 02 67
30 53 22 17 04	10 27 41 22 02	39 68 52 33 09	10 06 16 88 29	55 98 66 64 85
03 78 89 75 99	75 86 72 07 17	74 41 65 31 66	35 20 83 33 74	87 53 90 88 23
48 22 86 33 79	85 78 34 76 19	53 15 26 74 33	35 66 35 29 72	16 81 86 03 11
60 36 59 46 53	35 07 53 39 49	42 61 42 92 97	01 91 82 83 16	98 95 37 32 31
83 79 94 24 02	56 62 33 44 42	34 99 44 13 74	70 07 11 47 36	09 95 81 80 65
32 96 00 74 05	36 40 98 32 32	99 38 54 16 00	11 13 30 75 86	15 91 70 62 53
19 32 25 38 45	57 62 05 26 06	66 49 76 86 46	78 13 86 65 59	19 64 09 94 13
11 22 09 47 47	07 39 93 74 08	48 50 92 39 29	27 48 24 54 76	85 24 43 51 59
31 75 15 72 60	68 98 00 53 39	15 47 04 83 55	88 65 12 25 96	03 15 21 91 21
88 49 29 93 82	14 45 40 45 04	20 09 49 89 77	74 84 39 34 13	22 10 97 85 08
30 93 44 77 44	07 48 18 38 28	73 78 80 65 33	28 59 72 04 05	94 20 52 03 80
22 88 84 88 93	27 49 99 87 48	60 53 04 51 28	74 02 28 46 17	82 03 71 02 68
78 21 21 69 93	35 90 29 13 86	44 37 21 54 86	65 74 11 40 14	87 48 13 72 20
41 84 98 45 47	46 85 05 23 26	34 67 75 83 00	74 91 06 43 45	19 32 58 15 49
46 35 23 30 49	69 24 89 34 60	45 30 50 75 21	61 31 83 18 55	14 41 37 09 51
11 08 79 62 94	14 01 33 17 92	59 74 76 72 77	76 50 33 45 13	39 66 37 75 44
52 70 10 83 37	56 30 38 73 15	16 52 06 96 76	11 65 49 98 93	02 18 16 81 61
57 27 53 68 98	81 30 44 85 85	68 65 22 73 76	92 85 25 58 66	88 44 80 35 84
20 85 77 31 56	70 28 42 43 26	79 37 59 52 20	01 15 96 32 67	10 62 24 83 91
15 63 38 49 24	90 41 59 36 14	33 52 12 66 65	55 82 34 76 41	86 22 53 17 04
92 69 44 82 97	39 90 40 21 15	59 58 94 90 67	66 82 14 15 75	49 76 70 40 37
77 61 31 90 19	88 15 20 00 80	20 55 49 14 09	96 27 74 82 57	50 81 60 76 16
38 68 83 24 86	45 13 46 35 45	59 40 47 20 59	43 94 75 16 80	43 85 25 96 93
25 16 30 15 89	70 01 41 50 21	41 29 06 73 12	71 85 71 59 57	68 97 11 14 03
65 25 10 76 29	37 23 93 32 95	05 87 00 11 19	92 78 42 63 40	18 47 76 56 22
36 81 54 36 25	18 63 73 75 09	82 44 49 90 05	04 92 17 37 01	14 70 79 39 97
64 39 71 16 92	05 32 78 21 62	20 24 78 17 59	45 19 72 53 32	83 74 52 25 67
04 51 52 56 24	95 09 66 79 46	48 46 08 55 58	15 19 11 87 82	16 93 03 33 61
83 76 16 08 73	43 25 38 41 45	60 83 32 59 83	01 29 14 13 49	20 36 80 71 26
14 38 70 63 45	80 85 40 92 79	43 52 90 63 18	38 38 47 47 61	41 19 63 74 80
51 32 19 22 46	80 08 87 70 74	88 72 25 67 36	66 16 44 94 31	66 91 93 16 78
72 47 20 00 08	80 89 01 80 02	94 81 33 19 00	54 15 58 34 36	35 35 25 41 31
05 46 65 53 06	93 12 81 84 64	74 45 79 05 61	72 84 81 18 34	79 98 26 84 16
39 52 87 24 84	82 47 42 55 93	48 54 53 52 47	18 61 91 36 74	18 61 11 92 41
81 61 61 87 11	53 34 24 42 76	75 12 21 17 24	74 62 77 37 07	58 31 91 59 97
07 58 61 61 20	82 64 12 28 20	92 90 41 31 41	32 39 21 97 63	61 19 96 79 40
90 76 70 42 35	13 57 41 72 00	69 90 26 37 42	78 46 42 25 01	18 62 79 08 72
40 18 82 81 93	29 59 38 86 27	94 97 21 15 98	62 09 53 67 87	00 44 15 89 97
34 41 48 21 57	86 88 75 50 87	19 15 20 00 23	12 30 28 07 83	32 62 46 86 91
63 43 97 53 63	44 98 91 68 22	36 02 40 08 67	76 37 84 16 05	65 96 17 34 88
67 04 90 90 70	93 39 94 55 47	94 45 87 42 84	05 04 14 98 07	20 28 83 40 60
79 49 50 41 46	52 16 29 02 86	54 15 83 42 43	46 97 83 54 82	59 36 29 59 38
91 70 43 05 52	04 73 72 10 31	75 05 19 30 29	47 66 56 43 82	99 78 29 34 78

Source: Reproduced from Table A-1 of Wilfrid J. Dixon and Frank J. Massey, Jr., *Introduction to Statistical Analysis*, 3rd ed., McGraw-Hill Book Co., New York, 1969, with the permission of the RAND Corporation.

Table A.2. The Standard Normal Distribution[a]

$z[\lambda]$	λ	$z[\lambda]$	λ	$z[\lambda]$	λ	$z[\lambda]$	λ
.00	.5000						
.01	.5040	.26	.6026	.51	.6950	.76	.7764
.02	.5080	.27	.6064	.52	.6985	.77	.7794
.03	.5120	.28	.6103	.53	.7019	.78	.7823
.04	.5160	.29	.6141	.54	.7054	.79	.7852
.05	.5199	.30	.6179	.55	.7088	.80	.7881
.06	.5239	.31	.6217	.56	.7123	.81	.7910
.07	.5279	.32	.6255	.57	.7157	.82	.7939
.08	.5319	.33	.6293	.58	.7190	.83	.7967
.09	.5359	.34	.6331	.59	.7224	.84	.7995
.10	.5398	.35	.6368	.60	.7257	.85	.8023
.11	.5438	.36	.6406	.61	.7291	.86	.8051
.12	.5478	.37	.6443	.62	.7324	.87	.8078
.13	.5517	.38	.6480	.63	.7357	.88	.8106
.14	.5557	.39	.6517	.64	.7389	.89	.8133
.15	.5596	.40	.6554	.65	.7422	.90	.8159
.16	.5636	.41	.6591	.66	.7454	.91	.8186
.17	.5675	.42	.6628	.67	.7486	.92	.8212
.18	.5714	.43	.6664	.68	.7517	.93	.8238
.19	.5753	.44	.6700	.69	.7549	.94	.8264
.20	.5793	.45	.6736	.70	.7580	.95	.8289
.21	.5832	.46	.6772	.71	.7611	.96	.8315
.22	.5871	.47	.6808	.72	.7642	.97	.8340
.23	.5910	.48	.6844	.73	.7673	.98	.8365
.24	.5948	.49	.6879	.74	.7704	.99	.8389
.25	.5987	.50	.6915	.75	.7734	1.00	.8413

Table A.2. (*Continued*)

$z[\lambda]$	λ	$z[\lambda]$	λ	$z[\lambda]$	λ	$z[\lambda]$	λ
1.01	.8438	1.26	.8962	1.51	.9345	1.76	.9608
1.02	.8461	1.27	.8980	1.52	.9357	1.77	.9616
1.03	.8485	1.28	.8997	1.53	.9370	1.78	.9625
1.04	.8508	1.29	.9015	1.54	.9382	1.79	.9633
1.05	.8531	1.30	.9032	1.55	.9394	1.80	.9641
1.06	.8554	1.31	.9049	1.56	.9406	1.81	.9649
1.07	.8577	1.32	.9066	1.57	.9418	1.82	.9656
1.08	.8599	1.33	.9082	1.58	.9429	1.83	.9664
1.09	.8621	1.34	.9099	1.59	.9441	1.84	.9671
1.10	.8643	1.35	.9115	1.60	.9452	1.85	.9678
1.11	.8665	1.36	.9131	1.61	.9463	1.86	.9686
1.12	.8686	1.37	.9147	1.62	.9474	1.87	.9693
1.13	.8708	1.38	.9162	1.63	.9484	1.88	.9699
1.14	.8729	1.39	.9177	1.64	.9495	1.89	.9706
1.15	.8749	1.40	.9192	1.65	.9505	1.90	.9713
1.16	.8770	1.41	.9207	1.66	.9515	1.91	.9719
1.17	.8790	1.42	.9222	1.67	.9525	1.92	.9726
1.18	.8810	1.43	.9236	1.68	.9535	1.93	.9732
1.19	.8830	1.44	.9251	1.69	.9545	1.94	.9738
1.20	.8849	1.45	.9265	1.70	.9554	1.95	.9744
1.21	.8869	1.46	.9279	1.71	.9564	1.96	.9750
1.22	.8888	1.47	.9292	1.72	.9573	1.97	.9756
1.23	.8907	1.48	.9306	1.73	.9582	1.98	.9761
1.24	.8925	1.49	.9319	1.74	.9591	1.99	.9767
1.25	.8944	1.50	.9332	1.75	.9599	2.00	.9772

Table A.2. (*Continued*)

$z[\lambda]$	λ	$z[\lambda]$	λ	$z[\lambda]$	λ	$z[\lambda]$	λ
2.01	.9778	2.26	.9881	2.51	.9940	2.76	.9971
2.02	.9783	2.27	.9884	2.52	.9941	2.77	.9972
2.03	.9788	2.28	.9887	2.53	.9943	2.78	.9973
2.04	.9793	2.29	.9890	2.54	.9945	2.79	.9974
2.05	.9798	2.30	.9893	2.55	.9946	2.80	.9974
2.06	.9803	2.31	.9896	2.56	.9948	2.81	.9975
2.07	.9808	2.32	.9898	2.57	.9949	2.82	.9976
2.08	.9812	2.33	.9901	2.58	.9951	2.83	.9977
2.09	.9817	2.34	.9904	2.59	.9952	2.84	.9977
2.10	.9821	2.35	.9906	2.60	.9953	2.85	.9978
2.11	.9826	2.36	.9909	2.61	.9955	2.86	.9979
2.12	.9830	2.37	.9911	2.62	.9956	2.87	.9979
2.13	.9834	2.38	.9913	2.63	.9957	2.88	.9980
2.14	.9838	2.39	.9916	2.64	.9959	2.89	.9981
2.15	.9842	2.40	.9918	2.65	.9960	2.90	.9981
2.16	.9846	2.41	.9920	2.66	.9961	2.91	.9982
2.17	.9850	2.42	.9922	2.67	.9962	2.92	.9982
2.18	.9854	2.43	.9925	2.68	.9963	2.93	.9983
2.19	.9857	2.44	.9927	2.69	.9964	2.94	.9984
2.20	.9861	2.45	.9929	2.70	.9965	2.95	.9984
2.21	.9864	2.46	.9931	2.71	.9966	2.96	.9985
2.22	.9868	2.47	.9932	2.72	.9967	2.97	.9985
2.23	.9871	2.48	.9934	2.73	.9968	2.98	.9986
2.24	.9875	2.49	.9936	2.74	.9969	2.99	.9986
2.25	.9878	2.50	.9938	2.75	.9970	3.00	.9986

Table A.2. (*Continued*)

$z[\lambda]$	λ	$z[\lambda]$	λ	$z[\lambda]$	λ	$z[\lambda]$	λ
3.01	.9987	3.26	.9994	3.51	.9998	3.76	.9999
3.02	.9987	3.27	.9995	3.52	.9998	3.77	.9999
3.03	.9988	3.28	.9995	3.53	.9998	3.78	.9999
3.04	.9988	3.29	.9995	3.54	.9998	3.79	.9999
3.05	.9989	3.30	.9995	3.55	.9998	3.80	.9999
3.06	.9989	3.31	.9995	3.56	.9998	3.81	.9999
3.07	.9989	3.32	.9996	3.57	.9998	3.82	.9999
3.08	.9990	3.33	.9996	3.58	.9998	3.83	.9999
3.09	.9990	3.34	.9996	3.59	.9998	3.84	.9999
3.10	.9990	3.35	.9996	3.60	.9998	3.85	.9999
3.11	.9991	3.36	.9996	3.61	.9998	3.86	.9999
3.12	.9991	3.37	.9996	3.62	.9999	3.87	.9999
3.13	.9991	3.38	.9996	3.63	.9999	3.88	.9999
3.14	.9992	3.39	.9997	3.64	.9999	3.89	1.0000
3.15	.9992	3.40	.9997	3.65	.9999	3.90	1.0000
3.16	.9992	3.41	.9997	3.66	.9999	3.91	1.0000
3.17	.9992	3.42	.9997	3.67	.9999	3.92	1.0000
3.18	.9993	3.43	.9997	3.68	.9999	3.93	1.0000
3.19	.9993	3.44	.9997	3.69	.9999	3.94	1.0000
3.20	.9993	3.45	.9997	3.70	.9999	3.95	1.0000
3.21	.9993	3.46	.9997	3.71	.9999	3.96	1.0000
3.22	.9994	3.47	.9997	3.72	.9999	3.97	1.0000
3.23	.9994	3.48	.9997	3.73	.9999	3.98	1.0000
3.24	.9994	3.49	.9998	3.74	.9999	3.99	1.0000
3.25	.9994	3.50	.9998	3.75	.9999		

[a] λ = Area under curve from $-\infty$ to $z[\lambda]$.

$z[.50] = 0$, and for values of λ less than .50, $z[\lambda]$ is found by symmetry; for example, $z[.05] = -z[.95]$.
Source: Data in the table are extracted from Owen, D. B., *Handbook of Statistical Tables*, Addison-Wesley, Reading, MA., 1962. Courtesy of the U.S. Atomic Energy Commission.

Table A.3. The Students *t* Distribution[a]

d.f. \ λ	.75	.90	.95	.975	.990	.995	.999	.9995
1	1.000	3.078	6.314	12.706	31.821	63.657	318.309	636.619
2	.816	1.886	2.920	4.303	6.965	9.925	22.327	31.598
3	.765	1.638	2.353	3.182	4.541	5.841	10.214	12.924
4	.741	1.533	2.132	2.776	3.747	4.604	7.173	8.610
5	.727	1.476	2.015	2.571	3.365	4.032	5.893	6.869
6	.718	1.440	1.943	2.447	3.143	3.707	5.208	5.959
7	.711	1.415	1.895	2.365	2.998	3.499	4.785	5.408
8	.706	1.397	1.860	2.306	2.896	3.355	4.501	5.041
9	.703	1.383	1.833	2.262	2.821	3.250	4.297	4.781
10	.700	1.372	1.812	2.228	2.764	3.169	4.144	4.587
11	.697	1.363	1.796	2.201	2.718	3.106	4.025	4.437
12	.695	1.356	1.782	2.179	2.681	3.055	3.930	4.318
13	.694	1.350	1.771	2.160	2.650	3.012	3.852	4.221
14	.692	1.345	1.761	2.145	2.624	2.977	3.787	4.140
15	.691	1.341	1.753	2.131	2.602	2.947	3.733	4.073
16	.690	1.337	1.746	2.120	2.583	2.921	3.686	4.015
17	.689	1.333	1.740	2.110	2.567	2.898	3.646	3.965
18	.688	1.330	1.734	2.101	2.552	2.878	3.610	3.922
19	.688	1.328	1.729	2.093	2.539	2.861	3.579	3.883
20	.687	1.325	1.725	2.086	2.528	2.845	3.552	3.850
21	.686	1.323	1.721	2.080	2.518	2.831	3.527	3.819
22	.686	1.321	1.717	2.074	2.508	2.819	3.505	3.792
23	.685	1.319	1.714	2.069	2.500	2.807	3.485	3.768
24	.685	1.318	1.711	2.064	2.492	2.797	3.467	3.745
25	.684	1.316	1.708	2.060	2.485	2.787	3.450	3.725

Table A.3. (*Continued*)

d.f. \ λ	75	.90	.95	.975	.99	.995	.999	.9995
26	.684	1.315	1.706	2.056	2.479	2.779	3.435	3.707
27	.684	1.314	1.703	2.052	2.473	2.771	3.421	3.690
28	.683	1.313	1.701	2.048	2.467	2.763	3.408	3.674
29	.683	1.311	1.699	2.045	2.462	2.756	3.396	3.659
30	.683	1.310	1.697	2.042	2.457	2.750	3.385	3.646
35	.682	1.306	1.690	2.030	2.438	2.724	3.340	3.591
40	.681	1.303	1.684	2.021	2.423	2.704	3.307	3.551
45	.680	1.301	1.679	2.014	2.412	2.690	3.281	3.520
50	.679	1.299	1.676	2.009	2.403	2.678	3.261	3.496
55	.679	1.297	1.673	2.004	2.396	2.668	3.245	3.476
60	.679	1.296	1.671	2.000	2.390	2.660	3.232	3.460
70	.678	1.294	1.667	1.994	2.381	2.648	3.211	3.435
80	.678	1.292	1.664	1.990	2.374	2.639	3.195	3.416
90	.677	1.291	1.662	1.987	2.368	2.632	3.183	3.402
100	.677	1.290	1.660	1.984	2.364	2.626	3.174	3.390
120	.677	1.289	1.657	1.980	2.351	2.618	3.153	3.373
200	.676	1.286	1.652	1.972	2.345	2.601	3.131	3.340
500	.675	1.283	1.648	1.965	2.334	2.586	3.107	3.310
∞	.674	1.282	1.645	1.960	2.326	2.576	3.090	3.291

[a] λ = Area under curve from ∞ to $t[\lambda]$.

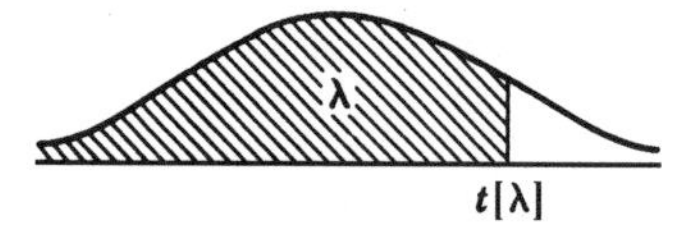

$t[.50] = 0$, and for values of λ less than .50, $t[\lambda]$ is found by symmetry: $t[\lambda] = -t[1 - \lambda]$. For example, $t[.025] = -t[.975]$. Note also that for d.f. $= \infty$, $t[\lambda] = z[\lambda]$.

Source: The data in this table are reprinted from E. T. Federighi, Extended Tables of the Percentage Points of Student's t-Distribution, *J. Am. Stat. Assoc.*, Sept. 1959, with the kind permission of the author.

Table A.4. The χ^2 Distribution[a]

d.f.[b] \ λ	.005	.01	.025	.05	.10	.90	.95	.975	.99	.995
1	.000039	.00016	.00098	.0039	.0158	2.71	3.84	5.02	6.63	7.88
2	.0100	.0201	.0506	.1026	.2107	4.61	5.99	7.38	9.21	10.60
3	.0717	.115	.216	.352	.594	6.25	7.81	9.35	11.34	12.84
4	.207	.297	.484	.711	1.064	7.78	9.49	11.14	13.28	14.86
5	.412	.554	.831	1.15	1.61	9.24	11.07	12.83	15.09	16.75
6	.676	.872	1.24	1.64	2.20	10.64	12.59	14.45	16.81	18.55
7	.989	1.24	1.69	2.17	2.83	12.02	14.07	16.01	18.48	20.28
8	1.34	1.65	2.18	2.73	3.49	13.36	15.51	17.53	20.09	21.96
9	1.73	2.09	2.70	3.33	4.17	14.68	16.92	19.02	21.67	23.59
10	2.16	2.56	3.25	3.94	4.87	15.99	18.31	20.48	23.21	25.19
11	2.60	3.05	3.82	4.57	5.58	17.28	19.68	21.92	24.73	26.76
12	3.07	3.57	4.40	5.23	6.30	18.55	21.03	23.34	26.22	28.30
13	3.57	4.11	5.01	5.89	7.04	19.81	22.36	24.74	27.69	29.82
14	4.07	4.66	5.63	6.57	7.79	21.06	23.68	26.12	29.14	31.32
15	4.60	5.23	6.26	7.26	8.55	22.31	25.00	27.49	30.58	32.80
16	5.14	5.81	6.91	7.96	9.31	23.54	26.30	28.85	32.00	34.27
18	6.26	7.01	8.23	9.39	10.86	25.99	28.87	31.53	34.81	37.16
20	7.43	8.26	9.59	10.85	12.44	28.41	31.41	34.17	37.57	40.00
24	9.89	10.86	12.40	13.85	15.66	33.20	36.42	39.36	42.98	45.56
30	13.79	14.95	16.79	18.49	20.60	40.26	43.77	46.98	50.89	53.67
40	20.71	22.16	24.43	26.51	29.05	51.81	55.76	59.34	63.69	66.77
60	35.53	37.48	40.48	43.19	46.46	74.40	79.08	83.30	88.38	91.95
120	83.85	86.92	91.58	95.70	100.62	140.23	146.57	152.21	158.95	163.64

[a] λ = Area under curve from 0 to $\chi^2[\lambda]$.

[b] For large values of d.f., the approximate formula

$$\chi^2[\lambda] = \nu\left(1 - \frac{2}{9\nu} + z[\lambda]\sqrt{\frac{2}{9\nu}}\right)^3$$

where $z[\lambda]$ is the normal deviate and ν is the number of degrees of freedom, may be used. For example, $\chi^2[.99] = 60[1 - .00370 + 2.326(.06086)]^3 = 60(1.1379)^3 = 88.4$ for the 99th percentile for 60 d.f.

Source: Reproduced from Table A-6a of W. J. Dixon and F. J. Massey, Jr., *Introduction to Statistical Analysis*, 3rd ed., McGraw-Hill Book Co., New York, 1969, with the permission of the authors and the publishers.

Table A.5. The F Distribution[a,b]

ν_2[c]	λ \ ν_1	1	2	3	4	5	6	7	8	9	10	11	12
1	.025	$.0^2 15$	.026	.057	.082	.100	.113	.124	.132	.139	.144	.149	.153
	.95	161	200	216	225	230	234	237	239	241	242	243	244
	.975	648	800	864	900	922	937	948	957	963	969	973	977
	.99	405^1	500^1	540^1	562^1	576^1	586^1	593^1	598^1	602^1	606^1	608^1	611^1
2	.025	$.0^2 13$	.026	.062	.094	.119	.138	.153	.165	.175	.183	.190	.196
	.95	18.5	19.0	19.2	19.2	19.3	19.3	19.4	19.4	19.4	19.4	19.4	19.4
	.975	38.5	39.0	39.2	39.2	39.3	39.3	39.4	39.4	39.4	39.4	39.4	39.4
	.99	98.5	99.0	99.2	99.2	99.3	99.3	99.4	99.4	99.4	99.4	99.4	99.4
3	.025	$.0^2 12$	.026	.065	.100	.129	.152	.170	.185	.197	.207	.216	.224
	.95	10.1	9.55	9.28	9.12	9.01	8.94	8.89	8.85	8.81	8.79	8.76	8.74
	.975	17.4	16.0	15.4	15.1	14.9	14.7	14.6	14.5	14.5	14.4	14.4	14.3
	.99	34.1	30.8	29.5	28.7	28.2	27.9	27.7	27.5	27.3	27.2	27.1	27.1
4	.025	$.0^2 11$	.026	.066	.104	.135	.161	.181	.198	.212	.224	.234	.243
	.95	7.71	6.94	6.59	6.39	6.26	6.16	6.09	6.04	6.00	5.96	5.94	5.91
	.975	12.2	10.6	9.98	9.60	9.36	9.20	9.07	8.98	8.90	8.84	8.79	8.75
	.99	21.2	18.0	16.7	16.0	15.5	15.2	15.0	14.8	14.7	14.5	14.4	14.4
5	.025	$.0^2 11$	.025	.067	.107	.140	.167	.189	.208	.223	.236	.248	.257
	.95	6.61	5.79	5.41	5.19	5.05	4.95	4.88	4.82	4.77	4.74	4.71	4.68
	.975	10.0	8.43	7.76	7.39	7.15	6.98	6.85	6.76	6.68	6.62	6.57	6.52
	.99	16.3	13.3	12.1	11.4	11.0	10.7	10.5	10.3	10.2	10.1	9.96	9.89
6	.025	$.0^2 11$	.025	.068	.109	.143	.172	.195	.215	.231	.246	.258	.268
	.95	5.99	5.14	4.76	4.53	4.39	4.28	4.21	4.15	4.10	4.06	4.03	4.00
	.975	8.81	7.26	6.60	6.23	5.99	5.82	5.70	5.60	5.52	5.46	5.41	5.37
	.99	13.7	10.9	9.78	9.15	8.75	8.47	8.26	8.10	7.98	7.87	7.79	7.72
7	.025	$.0^2 10$	.025	.068	.110	.146	.176	.200	.221	.238	.253	.266	.277
	.95	5.59	4.74	4.35	4.12	3.97	3.87	3.79	3.73	3.68	3.64	3.60	3.57
	.975	8.07	6.54	5.89	5.52	5.29	5.12	4.99	4.90	4.82	4.76	4.71	4.67
	.99	12.2	9.55	8.45	7.85	7.46	7.19	6.99	6.84	6.72	6.62	6.54	6.47
8	.025	$.0^2 10$	.025	.069	.111	.148	.179	.204	.226	.244	.259	.273	.285
	.95	5.32	4.46	4.07	3.84	3.69	3.58	3.50	3.44	3.39	3.35	3.31	3.28
	.975	7.57	6.06	5.42	5.05	4.82	4.65	4.53	4.43	4.36	4.30	4.24	4.20
	.99	11.3	8.65	7.59	7.01	6.63	6.37	6.18	6.03	5.91	5.81	5.73	5.67
9	.025	$.0^2 10$	.025	.069	.112	.150	.181	.207	.230	.248	.265	.279	.291
	.95	5.12	4.26	3.86	3.63	3.48	3.37	3.29	3.23	3.18	3.14	3.10	3.07
	.975	7.21	5.71	5.08	4.72	4.48	4.32	4.20	4.10	4.03	3.96	3.91	3.87
	.99	10.6	8.02	6.99	6.42	6.06	5.80	5.61	5.47	5.35	5.26	5.18	5.11
10	.025	$.0^2 10$	.025	.069	.113	.151	.183	.210	.233	.252	.269	.283	.296
	.95	4.96	4.10	3.71	3.48	3.33	3.22	3.14	3.07	3.02	2.98	2.94	2.91
	.975	6.94	5.46	4.83	4.47	4.24	4.07	3.95	3.85	3.78	3.72	3.66	3.62
	.99	10.0	7.56	6.55	5.99	5.64	5.39	5.20	5.06	4.94	4.85	4.77	4.71

Table A.5. ***(Continued)***

ν_2	λ \ ν_1	15	20	24	30	40	50	60	100	120	200	500	∞
1	.025	.161	.170	.175	.180	.184	.187	.189	.193	.194	.196	.198	.199
	.95	246	248	249	250	251	252	252	253	253	254	254	254
	.975	985	993	997	100^1	101^1	101^1	101^1	101^1	101^1	102^1	102^1	102^1
	.99	616^1	621^1	623^1	626^1	629^1	630^1	631^1	633^1	634^1	635^1	636^1	637^1
2	.025	.210	.224	.232	.239	.247	.251	.255	.261	.263	.266	.269	.271
	.95	19.4	19.4	19.5	19.5	19.5	19.5	19.5	19.5	19.5	19.5	19.5	19.5
	.975	39.4	39.4	39.5	39.5	39.5	39.5	39.5	39.5	39.5	39.5	39.5	39.5
	.99	99.4	99.4	99.5	99.5	99.5	99.5	99.5	99.5	99.5	99.5	99.5	99.5
3	.025	.241	.259	.269	.279	.289	.295	.299	.308	.310	.314	.318	.321
	.95	8.70	8.66	8.63	8.62	8.59	8.58	8.57	8.55	8.55	8.54	8.53	8.53
	.975	14.3	14.2	14.1	14.1	14.0	14.0	14.0	14.0	13.9	13.9	13.9	13.9
	.99	26.9	26.7	26.6	26.5	26.4	26.4	26.3	26.2	26.2	26.2	26.1	26.1
4	.025	.263	.284	.296	.308	.320	.327	.332	.342	.346	.351	.356	.359
	.95	5.86	5.80	5.77	5.75	5.72	5.70	5.69	5.66	5.66	5.65	5.64	5.63
	.975	8.66	8.56	8.51	8.46	8.41	8.38	8.36	8.32	8.31	8.29	8.27	8.26
	.99	14.2	14.0	13.9	13.8	13.7	13.7	13.7	13.6	13.6	13.5	13.5	13.5
5	.025	.280	.304	.317	.330	.344	.353	.359	.370	.374	.380	.386	.390
	.95	4.62	4.56	4.53	4.50	4.46	4.44	4.43	4.41	4.40	4.39	4.37	4.36
	.975	6.43	6.33	6.28	6.23	6.18	6.14	6.12	6.08	6.07	6.05	6.03	6.02
	.99	9.72	9.55	9.47	9.38	9.29	9.24	9.20	9.13	9.11	9.08	9.04	9.02
6	.025	.293	.320	.334	.349	.364	.375	.381	.394	.398	.405	.412	.415
	.95	3.94	3.87	3.84	3.81	3.77	3.75	3.74	3.71	3.70	3.69	3.68	3.67
	.975	5.27	5.17	5.12	5.07	5.01	4.98	4.96	4.92	4.90	4.88	4.86	4.85
	.99	7.56	7.40	7.31	7.23	7.14	7.09	7.06	6.99	6.97	6.93	6.90	6.88
7	.025	.304	.333	.348	.364	.381	.392	.399	.413	.418	.426	.433	.437
	.95	3.51	3.44	3.41	3.38	3.34	3.32	3.30	3.27	3.27	3.25	3.24	3.23
	.975	4.57	4.47	4.42	4.36	4.31	4.28	4.25	4.21	4.20	4.18	4.16	4.14
	.99	6.31	6.16	6.07	5.99	5.91	5.86	5.82	5.75	5.74	5.70	5.67	5.65
8	.025	.313	.343	.360	.377	.395	.407	.415	.431	.435	.442	.450	.456
	.95	3.22	3.15	3.12	3.08	3.04	3.02	3.01	2.97	2.97	2.95	2.94	2.93
	.975	4.10	4.00	3.95	3.89	3.84	3.81	3.78	3.74	3.73	3.70	3.68	3.67
	.99	5.52	5.36	5.28	5.20	5.12	5.07	5.03	4.96	4.95	4.91	4.88	4.86
9	.025	.320	.352	.370	.388	.408	.420	.428	.446	.450	.459	.467	.473
	.95	3.01	2.94	2.90	2.86	2.83	2.80	2.79	2.76	2.75	2.73	2.72	2.71
	.975	3.77	3.67	3.61	3.56	3.51	3.47	3.45	3.40	3.39	3.37	3.35	3.33
	.99	4.96	4.81	4.73	4.65	4.57	4.52	4.48	4.42	4.40	4.36	4.33	4.31
10	.025	.327	.360	.379	.398	.419	.431	.441	.459	.464	.474	.483	.488
	.95	2.85	2.77	2.74	2.70	2.66	2.64	2.62	2.59	2.58	2.56	2.55	2.54
	.975	3.52	3.42	3.37	3.31	3.26	3.22	3.20	3.15	3.14	3.12	3.09	3.08
	.99	4.56	4.41	4.33	4.25	4.17	4.12	4.08	4.01	4.00	3.96	3.93	3.91

Table A.5. ***(Continued)***

ν_2	λ \ ν_1	1	2	3	4	5	6	7	8	9	10	11	12
11	.025	$.0^{2}10$	.025	.069	.114	.152	.185	.212	.236	.256	.273	.288	.301
	.95	4.84	3.98	3.59	3.36	3.20	3.09	3.01	2.95	2.90	2.85	2.82	2.79
	.975	6.72	5.26	4.63	4.28	4.04	3.88	3.76	3.66	3.59	3.53	3.47	3.43
	.99	9.65	7.21	6.22	5.67	5.32	5.07	4.89	4.74	4.63	4.54	4.46	4.40
12	.025	$.0^{2}10$	.025	.070	.114	.153	.186	.214	.238	.259	.276	.292	.305
	.95	4.75	3.89	3.49	3.26	3.11	3.00	2.91	2.85	2.80	2.75	2.72	2.69
	.975	6.55	5.10	4.47	4.12	3.89	3.73	3.61	3.51	3.44	3.37	3.32	3.28
	.99	9.33	6.93	5.95	5.41	5.06	4.82	4.64	4.50	4.39	4.30	4.22	4.16
15	.025	$.0^{2}10$	.025	.070	.116	.156	.190	.219	.244	.265	.284	.300	.315
	.95	4.54	3.68	3.29	3.06	2.90	2.79	2.71	2.64	2.59	2.54	2.51	2.48
	.975	6.20	4.76	4.15	3.80	3.58	3.41	3.29	3.20	3.12	3.06	3.01	2.96
	.99	8.68	6.36	5.42	4.89	4.56	4.32	4.14	4.00	3.89	3.80	3.73	3.67
20	.025	$.0^{2}10$	.025	.071	.117	.158	.193	.224	.250	.273	.292	.310	.325
	.95	4.35	3.49	3.10	2.87	2.71	2.60	2.51	2.45	2.39	2.35	2.31	2.28
	.975	5.87	4.46	3.86	3.51	3.29	3.13	3.01	2.91	2.84	2.77	2.72	2.68
	.99	8.10	5.85	4.94	4.43	4.10	3.87	3.70	3.56	3.46	3.37	3.29	3.23
24	.025	$.0^{2}10$	.025	.071	.117	.159	.195	.227	.253	.277	.297	.315	.331
	.95	4.26	3.40	3.01	2.78	2.62	2.51	2.42	2.36	2.30	2.25	2.21	2.18
	.975	5.72	4.32	3.72	3.38	3.15	2.99	2.87	2.78	2.70	2.64	2.59	2.54
	.99	7.82	5.61	4.72	4.22	3.90	3.67	3.50	3.36	3.26	3.17	3.09	3.03
30	.025	$.0^{2}10$	.025	.071	.118	.161	.197	.229	.257	.281	.302	.321	.337
	.95	4.17	3.32	2.92	2.69	2.53	2.42	2.33	2.27	2.21	2.16	2.13	2.09
	.975	5.57	4.18	3.59	3.25	3.03	2.87	2.75	2.65	2.57	2.51	2.46	2.41
	.99	7.56	5.39	4.51	4.02	3.70	3.47	3.30	3.17	3.07	2.98	2.91	2.84
40	.025	$.0^{3}99$	.025	.071	.119	.162	.199	.232	.260	.285	.307	.327	.344
	.95	4.08	3.23	2.84	2.61	2.45	2.34	2.25	2.18	2.12	2.08	2.04	2.00
	.975	5.42	4.05	3.46	3.13	2.90	2.74	2.62	2.53	2.45	2.39	2.33	2.29
	.99	7.31	5.18	4.31	3.83	3.51	3.29	3.12	2.99	2.89	2.80	2.73	2.66
60	.025	$.0^{3}99$	.025	.071	.120	.163	.202	.235	.264	.290	.313	.333	.351
	.95	4.00	3.15	2.76	2.53	2.37	2.25	2.17	2.10	2.04	1.99	1.95	1.92
	.975	5.29	3.93	3.34	3.01	2.79	2.63	2.51	2.41	2.33	2.27	2.22	2.17
	.99	7.08	4.98	4.13	3.65	3.34	3.12	2.95	2.82	2.72	2.63	2.56	2.50
120	.025	$.0^{3}99$	.025	.072,	.120	.165	.204	.238	.268	.295	.318	.340	.359
	.95	3.92	3.07	2.68	2.45	2.29	2.18	2.09	2.02	1.96	1.91	1.87	1.83
	.975	5.15	3.80	3.23	2.89	2.67	2.52	2.39	2.30	2.22	2.16	2.10	2.05
	.99	6.85	4.79	3.95	3.48	3.17	2.96	2.79	2.66	2.56	2.47	2.40	2.34
∞	.025	$.0^{3}98$	.025	.072	.121	.166	.206	.241	.272	.300	.325	.347	.367
	.95	3.84	3.00	2.60	2.37	2.21	2.10	2.01	1.94	1.88	1.83	1.79	1.75
	.975	5.02	3.69	3.12	2.79	2.57	2.41	2.29	2.19	2.11	2.05	1.99	1.94
	.99	6.63	4.61	3.78	3.32	3.02	2.80	2.64	2.51	2.41	2.32	2.25	2.18

Table A.5. (*Continued*)

ν_2	λ \ ν_1	15	20	24	30	40	50	60	100	120	200	500	∞
11	.025	.332	.368	.386	.407	.429	.442	.450	.472	.476	.485	.495	.503
	.95	2.72	2.65	2.61	2.57	2.53	2.51	2.49	2.46	2.45	2.43	2.42	2.40
	.975	3.33	3.23	3.17	3.12	3.06	3.03	3.00	2.96	2.94	2.92	2.90	2.88
	.99	4.25	4.10	4.02	3.94	3.86	3.81	3.78	3.71	3.69	3.66	3.62	3.60
12	.025	.337	.374	.394	.416	.437	.450	.461	.481	.487	.498	.508	.514
	.95	2.62	2.54	2.51	2.47	2.43	2.40	2.38	2.35	2.34	2.32	2.31	2.30
	.975	3.18	3.07	3.02	2.96	2.91	2.87	2.85	2.80	2.79	2.76	2.74	2.72
	.99	4:01	3.86	3.78	3.70	3.62	3.57	3.54	3.47	3.45	3.41	3.38	3.36
15	.025	.349	.389	.410	.433	.458	.474	.485	.508	.514	.526	.538	.546
	.95	2.40	2.33	2.39	2.25	2.20	2.18	2.16	2.12	2.11	2.10	2.08	2.07
	.975	2.86	2.76	2.70	2.64	2.59	2.55	2.52	2.47	2.46	2.44	2.41	2.40
	.99	3.52	3.37	3.29	3.21	3.13	3.08	3.05	2.98	2.96	2.92	2.89	2.87
20	.025	.363	.406	.430	.456	.484	.503	.514	.541	.548	.562	.575	.585
	.95	2.20	2.12	2.08	2.04	1.99	1.97	1.95	1.91	1.90	1.88	1.86	1.84
	.975	2.57	2.46	2.41	2.35	2.29	2.25	2.22	2.17	2.16	2.13	2.10	2.09
	.99	3.09	2.94	2.86	2.78	2.69	2.64	2.61	2.54	2.52	2.48	2.44	2.42
24	.025	.370	.415	.441	.468	.498	.518	.531	.562	.568	.585	.599	.610
	.95	2.11	2.03	1.98	1.94	1.89	1.86	1.84	1.80	1.79	1.77	1.75	1.73
	.975	2.44	2.33	2.27	2.21	2.15	2.11	2.08	2.02	2.01	1.98	1.95	1.94
	.99	2.89	2.74	2.66	2.58	2.49	2.44	2.40	2.33	2.31	2.27	2.24	2.21
30	.025	.378	.426	.453	.482	.515	.535	.551	.585	.592	.610	.625	.639
	.95	2.01	1.93	1.89	1.84	1.79	1.76	1.74	1.70	1.68	1.66	1.64	1.62
	.975	2.31	2.20	2.14	2.07	2.01	1.97	1.94	1.88	1.87	1.84	1.81	1.79
	.99	2.70	2.55	2.47	2.39	2.30	2.25	2.21	2.13	2.11	2.07	2.03	2.01
40	.025	.387	.437	.466	.498	.533	.556	.573	.610	.620	.641	.662	.674
	.95	1.92	1.84	1.79	1.74	1.69	1.66	1.64	1.59	1.58	1.55	1.53	1.51
	.975	2.18	2.07	2.01	1.94	1.88	1.83	1.80	1.74	1.72	1.69	1.66	1.64
	.99	2.52	2.37	2.29	2.20	2.11	2.06	2.02	1.94	1.92	1.87	1.83	1.80
60	.025	.396	.450	.481	.515	.555	.581	.600	.641	.654	.680	.704	.720
	.95	1.84	1.75	1.70	1.65	1.59	1.56	1.53	1.48	1.47	1.44	1.41	1.39
	.975	2.06	1.94	1.88	1.82	1.74	1.70	1.67	1.60	1.58	1.54	1.51	1.48
	.99	2.35	2.20	2.12	2.03	1.94	1.88	1.84	1.75	1.73	1.68	1.63	1.60
120	.025	.406	.464	.498	.536	.580	.611	.633	.684	.698	.729	.762	.789
	.95	1.75	1.66	1.61	1.55	1.50	1.46	1.43	1.37	1.35	1.32	1.28	1.25
	.975	1.95	1.82	1.76	1.69	1.61	1.56	1.53	1.45	1.43	1.39	1.34	1.31
	.99	2.19	2.03	1.95	1.86	1.76	1.70	1.66	1.56	1.53	1.48	1.42	1.38
∞	.025	.418	.480	.517	.560	.611	.645	.675	.741	.763	.813	.878	1.00
	.95	1.67	1.57	1.52	1.46	1.39	1.35	1.32	1.24	1.22	1.17	1.11	1.00
	.975	1.83	1.71	1.64	1.57	1.48	1.43	1.39	1.30	1.27	1.21	1.13	1.00
	.99	2.04	1.88	1.79	1.70	1.59	1.52	1.47	1.36	1.32	1.25	1.15	1.00

[a] λ = Area under curve from 0 to $F[\lambda]$. To obtain $F[.05]$ and $F[.01]$ for degrees of freedom ν_1 and ν_2, use $1/F[.95]$ and $1/F[.99]$, respectively, with degrees of freedom ν_2 and ν_1.

[b] Notation: $593^3 = 593 \times 10^3$.

$.0^2 11 = .11 \times 10^{-2}$

[c] ν_1 is the d.f. for the first variance and ν_2 is the d.f. for the second variance.

Source: The data in this table were extracted from W. J. Dixon and F. J. Massey, Jr., *Introduction to Statistical Analysis*, 3rd ed., McGraw-Hill, 1969. Used with permission of McGraw-Hill Book Co.

Table A.6. Confidence Intervals for the Correlation Coefficient

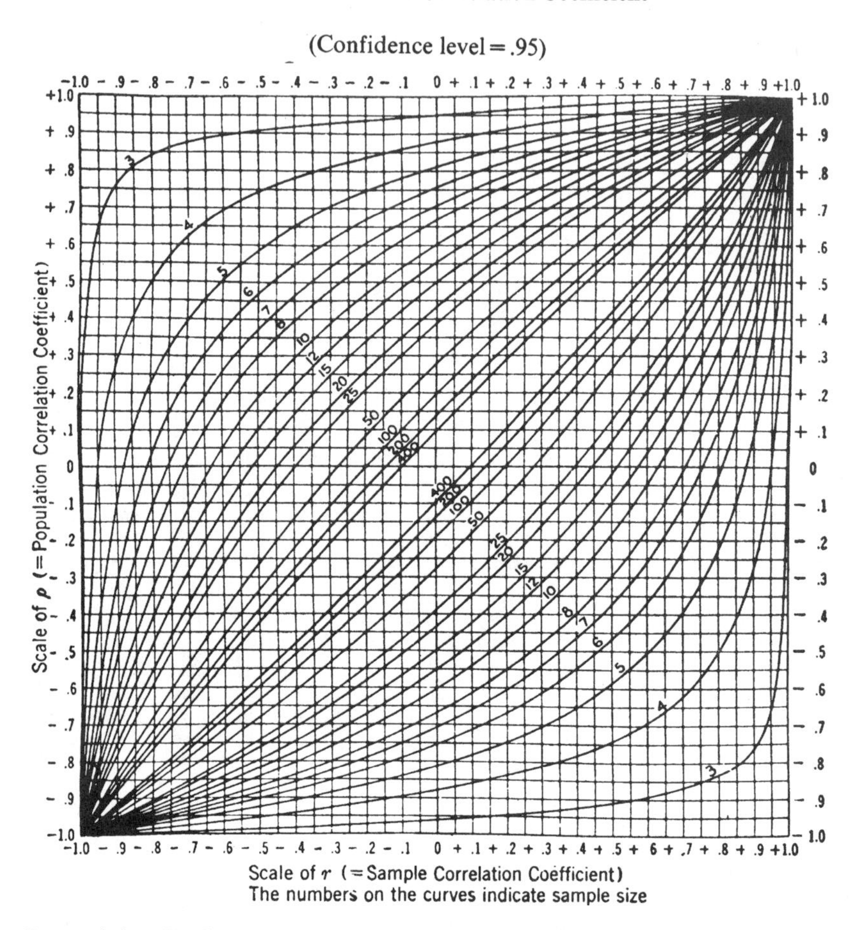

By permission of Prof. E.S. Pearson from F.N. David, *Tables of the Ordinates and Probablity Integral of the Distribution of the Correlation Coefficient in Small Sammples.* The Biometrika Office, London.

APPENDIX B

Answers to Selected Problems

Chapter 1

1.1 (a) Cross-sectional survey—asthma or osteoporosis in elderly; (b) case/control study—any cancer, rheumatoid arthritis; (c) a prospective panel study—emphysema in elderly heavy smokers, a second myocardial infarction; (d) clinical trial—myocardial infarction or melanoma.

1.2 No answer.

1.3 (a) High incidence rate—prospective study; (b) high prevalence—survey; (c) high funding—prospective study; (d) treatment under control of investigator—clinical trial.

1.4 Chronic diseases.

1.5 The purpose of the clinical trial is to determine if postmenopausal women who are given estrogen plus calcium supplements have a slower rate of decline in their bone mass than women given calcium supplements alone. Recommend clinical trial with random assignment and where healthy women who finished menopause within the last 6 months are followed for at least 5 years.

1.6 Case/control study.

Chapter 2

2.1 No answer.

2.2 No answer.

2.3 It may not appear to be random since there are more odd numbers than even numbers and there were six odd numbers in a row. But the numbers in this table are in a random order and this table has been carefully tested. What counts is how the numbers were generated. No, you will not have equal numbers in the two groups if 50 is the total sample size. Often when random assignment is used the sample size in the two treatment groups does not come out the same unless a method of forcing the sample size to be equal is used.

2.4 The easiest to explain is a systematic sample. We have $k = 180/25 = 7.2$, so pick a random number between 1 and 7, say 5. The instructions are starting with the fifth student in the list, take every seventh student from there on. The student numbers will be 5, 12, 19, 26, 33, and so on. Note that $12 = 5 + 7$ and $19 = 12 + 7$. In other words, you keep adding 7 to the previous number.

2.5 No answer.

2.6 Would use a stratified sample. This would ensure having an adequate sample size in each class.

2.7 First, sort the records of the cases by age and obtain the records of the women who were < 50 years old at time of diagnosis. One possibility is to chose matched neighborhood controls. Each control would be within say 5 years of age of the case and live in the same neighborhood. Hospital or friend controls could also be used.

Chapter 3

3.1 The stem and leaf graph can be obtained from the following compilations:

Stem	Leaf	Stem	Leaf
19	058	29	12444
20	5	30	0156
21	1	31	9
22	489	32	7
23		33	79
24	1199	34	8
25	0113367	35	
26	038	36	8
27	58	37	28
28	02347789	38	5

There was no excess of 0's or 5's in the leaf columns.

3.2–3.7 No answers.

3.8 Expect the vertical bars to have a similar height.

3.9 No answer.

Chapter 4

4.1 (a) Mean is 136.98 and the standard deviation is 32.80. (b) Median is 128.5 and $Q_1 = 114$ and $Q_3 = 154.25$. (c) Lower fourth is 114 and upper fourth is 153.5. (e) The mean is larger than the median and the difference between Q_1 and the median is 14.5, while the difference between Q_3 and the median is 25.75. The distribution appears to be nonsymmetric. (f) Check 87, 93, 212, and 230.

4.2 The mean is 14.0 and the standard deviation is 1.21.

4.3 The mean of all sample means is 3 and the mean of all sample variances would be the population variance or 2.

4.4 No answer.

4.5 No answer.

4.6 Age is ratio, gender is nominal, previous smoking status is nominal, nursing wing is nominal, length of stay is ratio, and four point scale is ordinal.

4.7 $\sum X^2 = X_1^2 + X_2^2 + X_3^2 + \cdots + X_n^2$ and $(\sum X)^2 = (X_1 + X_2 + X_3 + \cdots + X_n)^2$.

Chapter 5

5.1 Yes, the data appear to be symmetrically distributed and the distribution looks normal.

5.2 68.25% of the women.

5.3 Almost 100%.

5.4 15.87% of the women.

5.5 $P_{95} = 172.9$ pounds.

5.6 $P_{95} = 148.2$ pounds.

5.7 With a sample size of 16 and weights perhaps being a somewhat skewed distribution, we would expect the sample means to be approximately normally distributed.

5.8 Less than .005% or almost 0.

5.9 We are not concerned about normality since the question was about means and the sample size of 100 is large.

5.10 Yes, $z = 5$, so the P value is very small. Perhaps the sample was from a different population or was not a random sample. There is only a slight chance that it was simply random variation.

5.11 Likely transformations to try are a log or square root transformation since the distribution is skewed to the right.

Chapter 6

6.1 No answer.

6.2 No answer.

6.3 The 95% confidence limit is $1.04 < \mu < 2.09$. Since 1.407 is almost in the middle of the confidence limit, it appears that the pressure may not have changed the mean bleeding time.

6.4 The 95% confidence limit is $1.358 < \mu_1 - \mu_2 < 5.559$.

6.5 The 95% confidence limit is $.364 < \mu_1 - \mu_2 < .972$. Yes, since the confidence limit does not cover 0.

6.6 (a) No, the 95% confidence limit is $-.11$ to .99 kg. (b) If the sample size is 50, the 95% confidence limit is 0.13 to .75, so the answer is yes.

6.7 (a) The mean difference of $-.433$ degrees centigrade. (b) The 95% confidence interval is $-.7455$ to $-.1211$. In 95% of such experiments the computed intervals cover the true difference. Since the confidence limit does not include zero, the conclusion is that the rectal temperatures are higher than the oral ones. (c) The 95% confidence limit for the mean oral temperature is 37.3–37.9. (d) The 95% confidence limit for the mean rectal temperature is 37.6–38.5.

6.8 (a) $s_p = .0762$. (b) 18 d.f. (c) The 99% confidence interval is $-.410$ to $-.214$. (d) Yes, since the confidence limit does not include 0.

6.9 The 99% confidence limit is -50.12 to 2.12 with 11 d.f. A difference may exist but the data have failed to establish one since the interval covers 0.

6.10 The sample size in each group should be 217.

Chapter 7

7.1 No answer.

7.2 No answer.

7.3 $t = 4.38$ with 80 d.f. and $P < .0005$ so the null hypothesis is rejected. Pressure does appear to increase bleeding time.

7.4 $t = 3.62$ with 11 d.f. and $P = .004$ so reject null hypothesis. The male rats appear to be heavier.

7.5 $t = 9.158$ with 18 d.f. and $P < .0005$ so reject null hypothesis. Yes, it was effective.

7.6 (a) $t = -3.2004$ with 8 d.f. and $P = 0.013$. Conclude that the oral and rectal temperatures are different. (b) $t = 4.30$ with 8 d.f. and $P < .005$ so reject null hypothesis that the mean temperature is 37° C.

7.7 We do not have a random sample. We could correct for multiple testing using a Bonferroni correction.

7.8 The sample size should be 161 patients in each group.

7.9 $z = 1.56$, so we cannot reject the null hypothesis of equal means.

Chapter 8

8.1 Expect $\pi = .5$. The variance of an observation is $\pi(1 - \pi) = .25$. The variance of the mean of 10 observations is .025.

8.2 Only (g).

8.3 Expect 95.45%.

8.4 Expect 86.6%.

8.5 Less than .00005.

8.6 (a) The 95% confidence for the difference is $-.039$ to .239. There may be no difference between the remedies. (b) The sample size in the text was larger and the confidence interval computed there did not cover 0.

8.7 $z = 1.903$ so $P > .05$ and there is not a significant difference.

8.8 Need $n = 294$ if do not use the correction factor and 314 with the correction factor.

8.9 $z = 2.21$ and $P = .03$ so reject the null hypothesis. The 95% confidence interval is .013 to .167 and thus does not cover 0. The confidence interval gives us an interval that we are 95% certain will cover the true difference.

Chapter 9

9.1 (a) $\chi^2 = 8.22$ with 1 d.f. and $P = .004$ so reject the null hypothesis. (b) RR = 2.071. c) OR = 3.32. The 95% confidence limits of the odds ratio go from 1.44 to 7.65.

9.2 (a) OR = 2.33. (b) OR = 2.33. (c) χ^2 for (a) is 7.883 or $P = .0050$. For (b) χ^2 is 11.000 or $P = .0009$.

9.3 (a) The McNemar's χ^2 is 1.26 with 1 d.f. or nonsignificant. (b) The OR = 1.438 and the confidence interval goes from .732 to 2.721. Since the confidence limit covers 1, it also shows that the chemical was not shown to have an effect. In addition, the confidence limit has an approximately 95% chance of including the true odds ratio.

9.4 $\chi^2 = 23.27$ with 1 d.f. and $P < .0005$. OR = .26 of having been assaulted if currently living with spouse.

9.5 $\chi^2 = 5.084$ with 1 d.f. and $P = .0241$. Reject the null hypothesis.

9.6 $\chi^2 = 20.8232$ with 6 d.f. and $P = .002$. Reject null hypothesis of equal proportions.

9.10 Combine the results for moderate and severe heart failure.

Chapter 10

10.1 The 95% confidence limit for the population standard deviation goes from .312 to 1.228, so the null hypothesis is not rejected.

10.2 $F = 1.079$ with $\nu_1 = 9$ and $\nu_2 = 9$ d.f. Cannot reject the null hypothesis.

10.3 $F = 1.552$ with d.f. = 38, 42. Cannot reject null hypothesis.

10.4 The confidence limit is $.77 < \sigma_p^2 < 3.07$.

10.5 The data were taken from litter mates so the two samples are not independent.

10.6 The confidence limit goes from 2.94 to 60.35 for the variance of the difference with 7 d.f.

10.7 $F = 3.00$ with $\nu_1 = 20$ and $\nu_2 = 31$. Reject null hypothesis of equal variances. $t = 2.71$ with approximately 29 d.f. so reject null hypothesis of equal population means.

Chapter 11

11.1 (a) Fixed-X. (b) Rate $= -1518.8 + .7745$year. (c) The 95% confidence interval goes from .4637 to 1.0854. $t = 5.6360$ with 9 d.f. so $P = .0003$.

11.2 (a) Single sample. (b) Both negative. (c) Approximate 95% confidence limits are $-.92$ to $-.15$. (d) $t = -3.0026$ and $P = .0170$ in both cases.

11.3 (a) $r = .9236$. (b) From Problem 6.9, we could only conclude that the hypothesis of equal means could not be rejected. From the scatter diagram and $r = .9236$, we also learned that the two observers read each plate with a similar outcome.

11.4 (a) Single sample. (b) CBR $= 80.32 - .857$LifeExp. (c) Yes. (d) Decrease by 8.57. (e) $t = -6.651$ and $P < .00005$. Yes.

11.5 (a) LifeExp $= 63.48 + .0005$GDP. The line does not fit the points at all well. (b) $r = .71$. (c) The line fits much better after the transformation. $r = .88$. LifeExp $= 28.28 + 11.346 \log(\text{GDP})$. (d) The value of r is increased since r measures the linear association between X and Y.

11.6 (a) $r = -.7114$. CBR $= 26.6317 - .0005$GPD. (b) If use log(GDP), $r = -.8788$. The magnitude of r increased and the points fit a straight line much better. The correlation is a measure of the linear association between two variables.

11.7 (a) Originally $r = -.8642$ and $b = -.8570$. With an outlier in Y, $r = -.7517$ and $b = -.8676$. With an outlier in X, $r = -.8706$ and $b = -.8237$. With an outlier in both X and Y, $r = -.4818$ and $b = -.5115$. (d) The outlier in X and Y had by far the greater effect. The outlier in Y reduced r but did not change b hardly at all in this example.

Chapter 12

12.1 No answer.

12.2 Yes, the fit is quite good considering the small sample size.

12.3 The estimated median is .867.

12.4 Too small.

12.5 Kaplan–Meier method.

Days	Death	Censored	n_{obs}	$(n_{obs} - d)/n_{obs}$	$\hat{S}(t)$
15	1		7	(7-1)/7=.857	.857
18	2		6	(6-2)/6=.667	.572
21	1		4	(4-1)/4=.750	.429
21		1	3		
25		1	2		
26	1		1	(1-1)/1=0	.000

12.6 Clinical life table for pregnancy data.

Interval	n_{exp}	$\hat{q}$	$\hat{p}$	$\hat{S}(t)$	$\hat{h}(t)$
0 to < 1	371.5	.0108	.9892	1.0000	.0108
1 to < 2	347.5	.0086	.9914	.9892	.0087
2 to < 3	327.0	.0122	.9878	.9807	.0123
3 to < 4	305.5	.0098	.9902	.9687	.0099
4 to < 5	292.0	.0068	.9932	.9592	.0069
5 to < 6	283.0	.0141	.9859	.9526	.0142

Index

WILEY SERIES IN PROBABILITY AND STATISTICS
ESTABLISHED BY WALTER A. SHEWHART AND SAMUEL S. WILKS

Probability and Statistics Section

*ANDERSON · The Statistical Analysis of Time Series
ARNOLD, BALAKRISHNAN, and NAGARAJA · A First Course in Order Statistics
ARNOLD, BALAKRISHNAN, and NAGARAJA · Records
BACCELLI, COHEN, OLSDER, and QUADRAT · Synchronization and Linearity: An Algebra for Discrete Event Systems
BARNETT · Comparative Statistical Inference, *Third Edition*
BASILEVSKY · Statistical Factor Analysis and Related Methods: Theory and Applications
BERNARDO and SMITH · Bayesian Statistical Concepts and Theory
BILLINGSLEY · Convergence of Probability Measures, *Second Edition*
BOROVKOV · Asymptotic Methods in Queuing Theory
BOROVKOV · Ergodicity and Stability of Stochastic Processes
BRANDT, FRANKEN, and LISEK · Stationary Stochastic Models
CAINES · Linear Stochastic Systems
CAIROLI and DALANG · Sequential Stochastic Optimization
CONSTANTINE · Combinatorial Theory and Statistical Design
COOK · Regression Graphics
COVER and THOMAS · Elements of Information Theory
CSÖRGŐ and HORVÁTH · Weighted Approximations in Probability Statistics
CSÖRGŐ and HORVÁTH · Limit Theorems in Change Point Analysis
*DANIEL · Fitting Equations to Data: Computer Analysis of Multifactor Data, *Second Edition*
DETTE and STUDDEN · The Theory of Canonical Moments with Applications in Statistics, Probability, and Analysis
DEY and MUKERJEE · Fractional Factorial Plans
*DOOB · Stochastic Processes
DRYDEN and MARDIA · Statistical Shape Analysis
DUPUIS and ELLIS · A Weak Convergence Approach to the Theory of Large Deviations
ETHIER and KURTZ · Markov Processes: Characterization and Convergence
FELLER · An Introduction to Probability Theory and Its Applications, Volume I, *Third Edition,* Revised; Volume II, *Second Edition*
FULLER · Introduction to Statistical Time Series, *Second Edition*
FULLER · Measurement Error Models
GHOSH, MUKHOPADHYAY, and SEN · Sequential Estimation
GIFI · Nonlinear Multivariate Analysis
GUTTORP · Statistical Inference for Branching Processes
HALL · Introduction to the Theory of Coverage Processes
HAMPEL · Robust Statistics: The Approach Based on Influence Functions
HANNAN and DEISTLER · The Statistical Theory of Linear Systems
HUBER · Robust Statistics

*Now available in a lower priced paperback edition in the Wiley Classics Library.

Probability and Statistics (Continued)

HUSKOVA, BERAN, and DUPAC · Collected Works of Jaroslav Hajek—with Commentary
IMAN and CONOVER · A Modern Approach to Statistics
JUREK and MASON · Operator-Limit Distributions in Probability Theory
KASS and VOS · Geometrical Foundations of Asymptotic Inference
KAUFMAN and ROUSSEEUW · Finding Groups in Data: An Introduction to Cluster Analysis
KELLY · Probability, Statistics, and Optimization
KENDALL, BARDEN, CARNE, and LE · Shape and Shape Theory
LINDVALL · Lectures on the Coupling Method
McFADDEN · Management of Data in Clinical Trials
MANTON, WOODBURY, and TOLLEY · Statistical Applications Using Fuzzy Sets
MORGENTHALER and TUKEY · Configural Polysampling: A Route to Practical Robustness
MUIRHEAD · Aspects of Multivariate Statistical Theory
OLIVER and SMITH · Influence Diagrams, Belief Nets and Decision Analysis
*PARZEN · Modern Probability Theory and Its Applications
PEÑA, TIAO, and TSAY · A Course in Time Series Analysis
PRESS · Bayesian Statistics: Principles, Models, and Applications
PUKELSHEIM · Optimal Experimental Design
RAO · Asymptotic Theory of Statistical Inference
RAO · Linear Statistical Inference and Its Applications, *Second Edition*
RAO and SHANBHAG · Choquet-Deny Type Functional Equations with Applications to Stochastic Models
ROBERTSON, WRIGHT, and DYKSTRA · Order Restricted Statistical Inference
ROGERS and WILLIAMS · Diffusions, Markov Processes, and Martingales, Volume I: Foundations, *Second Edition;* Volume II: Îto Calculus
RUBINSTEIN and SHAPIRO · Discrete Event Systems: Sensitivity Analysis and Stochastic Optimization by the Score Function Method
RUZSA and SZEKELY · Algebraic Probability Theory
SCHEFFE · The Analysis of Variance
SEBER · Linear Regression Analysis
SEBER · Multivariate Observations
SEBER and WILD · Nonlinear Regression
SERFLING · Approximation Theorems of Mathematical Statistics
SHORACK and WELLNER · Empirical Processes with Applications to Statistics
SMALL and McLEISH · Hilbert Space Methods in Probability and Statistical Inference
STAPLETON · Linear Statistical Models
STAUDTE and SHEATHER · Robust Estimation and Testing
STOYANOV · Counterexamples in Probability
TANAKA · Time Series Analysis: Nonstationary and Noninvertible Distribution Theory
THOMPSON and SEBER · Adaptive Sampling
WELSH · Aspects of Statistical Inference
WHITTAKER · Graphical Models in Applied Multivariate Statistics
YANG · The Construction Theory of Denumerable Markov Processes

Applied Probability and Statistics Section

ABRAHAM and LEDOLTER · Statistical Methods for Forecasting
AGRESTI · Analysis of Ordinal Categorical Data

*Now available in a lower priced paperback edition in the Wiley Classics Library.

Applied Probability and Statistics (Continued)

AGRESTI · Categorical Data Analysis
ANDERSON, AUQUIER, HAUCK, OAKES, VANDAELE, and WEISBERG · Statistical Methods for Comparative Studies
ARMITAGE and DAVID (editors) · Advances in Biometry
*ARTHANARI and DODGE · Mathematical Programming in Statistics
ASMUSSEN · Applied Probability and Queues
*BAILEY · The Elements of Stochastic Processes with Applications to the Natural Sciences
BARNETT and LEWIS · Outliers in Statistical Data, *Third Edition*
BARTHOLOMEW, FORBES, and McLEAN · Statistical Techniques for Manpower Planning, *Second Edition*
BASU and RIGDON · Statistical Methods for the Reliability of Repairable Systems
BATES and WATTS · Nonlinear Regression Analysis and Its Applications
BECHHOFER, SANTNER, and GOLDSMAN · Design and Analysis of Experiments for Statistical Selection, Screening, and Multiple Comparisons
BELSLEY · Conditioning Diagnostics: Collinearity and Weak Data in Regression
BELSLEY, KUH, and WELSCH · Regression Diagnostics: Identifying Influential Data and Sources of Collinearity
BHAT · Elements of Applied Stochastic Processes, *Second Edition*
BHATTACHARYA and WAYMIRE · Stochastic Processes with Applications
BIRKES and DODGE · Alternative Methods of Regression
BLISCHKE AND MURTHY · Reliability: Modeling, Prediction, and Optimization
BLOOMFIELD · Fourier Analysis of Time Series: An Introduction, *Second Edition*
BOLLEN · Structural Equations with Latent Variables
BOULEAU · Numerical Methods for Stochastic Processes
BOX · Bayesian Inference in Statistical Analysis
BOX and DRAPER · Empirical Model-Building and Response Surfaces
*BOX and DRAPER · Evolutionary Operation: A Statistical Method for Process Improvement
BUCKLEW · Large Deviation Techniques in Decision, Simulation, and Estimation
BUNKE and BUNKE · Nonlinear Regression, Functional Relations and Robust Methods: Statistical Methods of Model Building
CHATTERJEE and HADI · Sensitivity Analysis in Linear Regression
CHERNICK · Bootstrap Methods: A Practitioner's Guide
CHILÈS and DELFINER · Geostatistics: Modeling Spatial Uncertainty
CHOW and LIU · Design and Analysis of Clinical Trials: Concepts and Methodologies
CLARKE and DISNEY · Probability and Random Processes: A First Course with Applications, *Second Edition*
*COCHRAN and COX · Experimental Designs, *Second Edition*
CONOVER · Practical Nonparametric Statistics, *Second Edition*
CORNELL · Experiments with Mixtures, Designs, Models, and the Analysis of Mixture Data, *Second Edition*
*COX · Planning of Experiments
CRESSIE · Statistics for Spatial Data, *Revised Edition*
DANIEL · Applications of Statistics to Industrial Experimentation
DANIEL · Biostatistics: A Foundation for Analysis in the Health Sciences, *Sixth Edition*
DAVID · Order Statistics, *Second Edition*
*DEGROOT, FIENBERG, and KADANE · Statistics and the Law
DODGE · Alternative Methods of Regression
DOWDY and WEARDEN · Statistics for Research, *Second Edition*
DUNN and CLARK · Applied Statistics: Analysis of Variance and Regression, *Second Edition*
*ELANDT-JOHNSON and JOHNSON · Survival Models and Data Analysis

*Now available in a lower priced paperback edition in the Wiley Classics Library.

Applied Probability and Statistics (Continued)

*FLEISS · The Design and Analysis of Clinical Experiments
FLEISS · Statistical Methods for Rates and Proportions, *Second Edition*
FLEMING and HARRINGTON · Counting Processes and Survival Analysis
GALLANT · Nonlinear Statistical Models
GLASSERMAN and YAO · Monotone Structure in Discrete-Event Systems
GNANADESIKAN · Methods for Statistical Data Analysis of Multivariate Observations, *Second Edition*
GOLDSTEIN and LEWIS · Assessment: Problems, Development, and Statistical Issues
GREENWOOD and NIKULIN · A Guide to Chi-Squared Testing
*HAHN · Statistical Models in Engineering
HAHN and MEEKER · Statistical Intervals: A Guide for Practitioners
HAND · Construction and Assessment of Classification Rules
HAND · Discrimination and Classification
HEIBERGER · Computation for the Analysis of Designed Experiments
HEDAYAT and SINHA · Design and Inference in Finite Population Sampling
HINKELMAN and KEMPTHORNE: · Design and Analysis of Experiments, Volume 1: Introduction to Experimental Design
HOAGLIN, MOSTELLER, and TUKEY · Exploratory Approach to Analysis of Variance
HOAGLIN, MOSTELLER, and TUKEY · Exploring Data Tables, Trends and Shapes
HOAGLIN, MOSTELLER, and TUKEY · Understanding Robust and Exploratory Data Analysis
HOCHBERG and TAMHANE · Multiple Comparison Procedures
HOCKING · Methods and Applications of Linear Models: Regression and the Analysis of Variables
HOGG and KLUGMAN · Loss Distributions
HOSMER and LEMESHOW · Applied Logistic Regression, *Second Edition*
HØYLAND and RAUSAND · System Reliability Theory: Models and Statistical Methods
HUBERTY · Applied Discriminant Analysis
HUNT and KENNEDY · Financial Derivatives in Theory and Practice
JACKSON · A User's Guide to Principle Components
JOHN · Statistical Methods in Engineering and Quality Assurance
JOHNSON · Multivariate Statistical Simulation
JOHNSON and KOTZ · Distributions in Statistics
JOHNSON, KOTZ, and BALAKRISHNAN · Continuous Univariate Distributions, Volume 1, *Second Edition*
JOHNSON, KOTZ, and BALAKRISHNAN · Continuous Univariate Distributions, Volume 2, *Second Edition*
JOHNSON, KOTZ, and BALAKRISHNAN · Discrete Multivariate Distributions
JOHNSON, KOTZ, and KEMP · Univariate Discrete Distributions, *Second Edition*
JUREČKOVÁ and SEN · Robust Statistical Procedures: Aymptotics and Interrelations
KADANE · Bayesian Methods and Ethics in a Clinical Trial Design
KADANE AND SCHUM · A Probabilistic Analysis of the Sacco and Vanzetti Evidence
KALBFLEISCH and PRENTICE · The Statistical Analysis of Failure Time Data
KELLY · Reversability and Stochastic Networks
KHURI, MATHEW, and SINHA · Statistical Tests for Mixed Linear Models
KLUGMAN, PANJER, and WILLMOT · Loss Models: From Data to Decisions
KLUGMAN, PANJER, and WILLMOT · Solutions Manual to Accompany Loss Models: From Data to Decisions
KOTZ, BALAKRISHNAN, and JOHNSON · Continuous Multivariate Distributions, Volume 1, *Second Edition*
KOVALENKO, KUZNETZOV, and PEGG · Mathematical Theory of Reliability of Time-Dependent Systems with Practical Applications

*Now available in a lower priced paperback edition in the Wiley Classics Library.

Applied Probability and Statistics (Continued)

LACHIN · Biostatistical Methods: The Assessment of Relative Risks
LAD · Operational Subjective Statistical Methods: A Mathematical, Philosophical, and Historical Introduction
LANGE, RYAN, BILLARD, BRILLINGER, CONQUEST, and GREENHOUSE · Case Studies in Biometry
LAWLESS · Statistical Models and Methods for Lifetime Data
LEE · Statistical Methods for Survival Data Analysis, *Second Edition*
LEPAGE and BILLARD · Exploring the Limits of Bootstrap
LINHART and ZUCCHINI · Model Selection
LITTLE and RUBIN · Statistical Analysis with Missing Data
LLOYD · The Statistical Analysis of Categorical Data
MAGNUS and NEUDECKER · Matrix Differential Calculus with Applications in Statistics and Econometrics, *Revised Edition*
MALLER and ZHOU · Survival Analysis with Long Term Survivors
MANN, SCHAFER, and SINGPURWALLA · Methods for Statistical Analysis of Reliability and Life Data
McLACHLAN · Discriminant Analysis and Statistical Pattern Recognition
McLACHLAN and KRISHNAN · The EM Algorithm and Extensions
McLACHLAN and PEEL · Finite Mixture Models
McNEIL · Epidemiological Research Methods
MEEKER and ESCOBAR · Statistical Methods for Reliability Data
*MILLER · Survival Analysis, *Second Edition*
MONTGOMERY and PECK · Introduction to Linear Regression Analysis, *Second Edition*
MYERS and MONTGOMERY · Response Surface Methodology: Process and Product in Optimization Using Designed Experiments
NELSON · Accelerated Testing, Statistical Models, Test Plans, and Data Analyses
NELSON · Applied Life Data Analysis
OCHI · Applied Probability and Stochastic Processes in Engineering and Physical Sciences
OKABE, BOOTS, and SUGIHARA · Spatial Tesselations: Concepts and Applications of Voronoi Diagrams
PANKRATZ · Forecasting with Dynamic Regression Models
PANKRATZ · Forecasting with Univariate Box-Jenkins Models: Concepts and Cases
PIANTADOSI · Clinical Trials: A Methodologic Perspective
PORT · Theoretical Probability for Applications
PUTERMAN · Markov Decision Processes: Discrete Stochastic Dynamic Programming
RACHEV · Probability Metrics and the Stability of Stochastic Models
RÉNYI · A Diary on Information Theory
RIPLEY · Spatial Statistics
RIPLEY · Stochastic Simulation
ROLSKI, SCHMIDLI, SCHMIDT, and TEUGELS · Stochastic Processes for Insurance and Finance
ROUSSEEUW and LEROY · Robust Regression and Outlier Detection
RUBIN · Multiple Imputation for Nonresponse in Surveys
RUBINSTEIN · Simulation and the Monte Carlo Method
RUBINSTEIN and MELAMED · Modern Simulation and Modeling
RYAN · Statistical Methods for Quality Improvement, *Second Edition*
SCHIMEK · Smoothing and Regression: Approaches, Computation, and Application
SCHUSS · Theory and Applications of Stochastic Differential Equations
SCOTT · Multivariate Density Estimation: Theory, Practice, and Visualization
*SEARLE · Linear Models
SEARLE · Linear Models for Unbalanced Data
SEARLE, CASELLA, and McCULLOCH · Variance Components

*Now available in a lower priced paperback edition in the Wiley Classics Library.

Applied Probability and Statistics (Continued)

SENNOTT · Stochastic Dynamic Programming and the Control of Queueing Systems
STOYAN, KENDALL, and MECKE · Stochastic Geometry and Its Applications, *Second Edition*
STOYAN and STOYAN · Fractals, Random Shapes and Point Fields: Methods of Geometrical Statistics
THOMPSON · Empirical Model Building
THOMPSON · Sampling
THOMPSON · Simulation: A Modeler's Approach
TIJMS · Stochastic Modeling and Analysis: A Computational Approach
TIJMS · Stochastic Models: An Algorithmic Approach
TITTERINGTON, SMITH, and MAKOV · Statistical Analysis of Finite Mixture Distributions
UPTON and FINGLETON · Spatial Data Analysis by Example, Volume 1: Point Pattern and Quantitative Data
UPTON and FINGLETON · Spatial Data Analysis by Example, Volume II: Categorical and Directional Data
VAN RIJCKEVORSEL and DE LEEUW · Component and Correspondence Analysis
VIDAKOVIC · Statistical Modeling by Wavelets
WEISBERG · Applied Linear Regression, *Second Edition*
WESTFALL and YOUNG · Resampling-Based Multiple Testing: Examples and Methods for *p*-Value Adjustment
WHITTLE · Systems in Stochastic Equilibrium
WOODING · Planning Pharmaceutical Clinical Trials: Basic Statistical Principles
WOOLSON · Statistical Methods for the Analysis of Biomedical Data
*ZELLNER · An Introduction to Bayesian Inference in Econometrics

Texts and References Section

AGRESTI · An Introduction to Categorical Data Analysis
ANDERSON · An Introduction to Multivariate Statistical Analysis, *Second Edition*
ANDERSON and LOYNES · The Teaching of Practical Statistics
ARMITAGE and COLTON · Encyclopedia of Biostatistics: Volumes 1 to 6 with Index
BARTOSZYNSKI and NIEWIADOMSKA-BUGAJ · Probability and Statistical Inference
BENDAT and PIERSOL · Random Data: Analysis and Measurement Procedures, *Third Edition*
BERRY, CHALONER, and GEWEKE · Bayesian Analysis in Statistics and Econometrics: Essays in Honor of Arnold Zellner
BHATTACHARYA and JOHNSON · Statistical Concepts and Methods
BILLINGSLEY · Probability and Measure, *Second Edition*
BOX · R. A. Fisher, the Life of a Scientist
BOX, HUNTER, and HUNTER · Statistics for Experimenters: An Introduction to Design, Data Analysis, and Model Building
BOX and LUCEÑO · Statistical Control by Monitoring and Feedback Adjustment
BROWN and HOLLANDER · Statistics: A Biomedical Introduction
CHATTERJEE and PRICE · Regression Analysis by Example, *Third Edition*
COOK and WEISBERG · Applied Regression Including Computing and Graphics
COOK and WEISBERG · An Introduction to Regression Graphics
COX · A Handbook of Introductory Statistical Methods
DILLON and GOLDSTEIN · Multivariate Analysis: Methods and Applications
*DODGE and ROMIG · Sampling Inspection Tables, *Second Edition*
DRAPER and SMITH · Applied Regression Analysis, *Third Edition*
DUDEWICZ and MISHRA · Modern Mathematical Statistics

*Now available in a lower priced paperback edition in the Wiley Classics Library.

Texts and References (Continued)

DUNN and CLARK · Basic Statistics: A Primer for the Biomedical Sciences, *Third Edition*

EVANS, HASTINGS, and PEACOCK · Statistical Distributions, *Third Edition*

FISHER and VAN BELLE · Biostatistics: A Methodology for the Health Sciences

FREEMAN and SMITH · Aspects of Uncertainty: A Tribute to D. V. Lindley

GROSS and HARRIS · Fundamentals of Queueing Theory, *Third Edition*

HALD · A History of Probability and Statistics and their Applications Before 1750

HALD · A History of Mathematical Statistics from 1750 to 1930

HELLER · MACSYMA for Statisticians

HOEL · Introduction to Mathematical Statistics, *Fifth Edition*

HOLLANDER and WOLFE · Nonparametric Statistical Methods, *Second Edition*

HOSMER and LEMESHOW · Applied Logistic Regression, *Second Edition*

HOSMER and LEMESHOW · Applied Survival Analysis: Regression Modeling of Time to Event Data

JOHNSON and BALAKRISHNAN · Advances in the Theory and Practice of Statistics: A Volume in Honor of Samuel Kotz

JOHNSON and KOTZ (editors) · Leading Personalities in Statistical Sciences: From the Seventeenth Century to the Present

JUDGE, GRIFFITHS, HILL, LÜTKEPOHL, and LEE · The Theory and Practice of Econometrics, *Second Edition*

KHURI · Advanced Calculus with Applications in Statistics

KOTZ and JOHNSON (editors) · Encyclopedia of Statistical Sciences: Volumes 1 to 9 with Index

KOTZ and JOHNSON (editors) · Encyclopedia of Statistical Sciences: Supplement Volume

KOTZ, REED, and BANKS (editors) · Encyclopedia of Statistical Sciences: Update Volume 1

KOTZ, REED, and BANKS (editors) · Encyclopedia of Statistical Sciences: Update Volume 2

LAMPERTI · Probability: A Survey of the Mathematical Theory, *Second Edition*

LARSON · Introduction to Probability Theory and Statistical Inference, *Third Edition*

LE · Applied Categorical Data Analysis

LE · Applied Survival Analysis

MALLOWS · Design, Data, and Analysis by Some Friends of Cuthbert Daniel

MARDIA · The Art of Statistical Science: A Tribute to G. S. Watson

MASON, GUNST, and HESS · Statistical Design and Analysis of Experiments with Applications to Engineering and Science

MURRAY · X-STAT 2.0 Statistical Experimentation, Design Data Analysis, and Nonlinear Optimization

PURI, VILAPLANA, and WERTZ · New Perspectives in Theoretical and Applied Statistics

RENCHER · Linear Models in Statistics

RENCHER · Methods of Multivariate Analysis

RENCHER · Multivariate Statistical Inference with Applications

ROSS · Introduction to Probability and Statistics for Engineers and Scientists

ROHATGI · An Introduction to Probability Theory and Mathematical Statistics

ROHATGI and SALEH · An Introduction to Probability Theory and Mathematical Statistics, *Second Edition*

RYAN · Modern Regression Methods

SCHOTT · Matrix Analysis for Statistics

SEARLE · Matrix Algebra Useful for Statistics

STYAN · The Collected Papers of T. W. Anderson: 1943–1985

TIAO, BISGAARD, HILL, PEÑA, and STIGLER (editors) · Box on Quality and Discovery: with Design, Control, and Robustness

*Now available in a lower priced paperback edition in the Wiley Classics Library.

Texts and References (Continued)

TIERNEY · LISP-STAT: An Object-Oriented Environment for Statistical Computing and Dynamic Graphics
WONNACOTT and WONNACOTT · Econometrics, *Second Edition*
WU and HAMADA · Experiments: Planning, Analysis, and Parameter Design Optimization

WILEY SERIES IN PROBABILITY AND STATISTICS

ESTABLISHED BY WALTER A. SHEWHART AND SAMUEL S. WILKS

Editors
Robert M. Groves, Graham Kalton, J. N. K. Rao, Norbert Schwarz, Christopher Skinner

Survey Methodology Section

BIEMER, GROVES, LYBERG, MATHIOWETZ, and SUDMAN · Measurement Errors in Surveys
COCHRAN · Sampling Techniques, *Third Edition*
COUPER, BAKER, BETHLEHEM, CLARK, MARTIN, NICHOLLS, and O'REILLY (editors) · Computer Assisted Survey Information Collection
COX, BINDER, CHINNAPPA, CHRISTIANSON, COLLEDGE, and KOTT (editors) · Business Survey Methods
*DEMING · Sample Design in Business Research
DILLMAN · Mail and Telephone Surveys: The Total Design Method, *Second Edition*
GROVES and COUPER · Nonresponse in Household Interview Surveys
GROVES · Survey Errors and Survey Costs
GROVES, BIEMER, LYBERG, MASSEY, NICHOLLS, and WAKSBERG · Telephone Survey Methodology
*HANSEN, HURWITZ, and MADOW · Sample Survey Methods and Theory, Volume 1: Methods and Applications
*HANSEN, HURWITZ, and MADOW · Sample Survey Methods and Theory, Volume II: Theory
KISH · Statistical Design for Research
*KISH · Survey Sampling
KORN and GRAUBARD · Analysis of Health Surveys
LESSLER and KALSBEEK · Nonsampling Error in Surveys
LEVY and LEMESHOW · Sampling of Populations: Methods and Applications, *Third Edition*
LYBERG, BIEMER, COLLINS, de LEEUW, DIPPO, SCHWARZ, TREWIN (editors) · Survey Measurement and Process Quality
SIRKEN, HERRMANN, SCHECHTER, SCHWARZ, TANUR, and TOURANGEAU (editors) · Cognition and Survey Research
VALLIANT, DORFMAN, and ROYALL · A Finite Population Sampling and Inference

*Now available in a lower priced paperback edition in the Wiley Classics Library.